INFERENCE, METHOD AND DECISION

SYNTHESE LIBRARY

MONOGRAPHS ON EPISTEMOLOGY,

LOGIC, METHODOLOGY, PHILOSOPHY OF SCIENCE,

SOCIOLOGY OF SCIENCE AND OF KNOWLEDGE,

AND ON THE MATHEMATICAL METHODS OF

SOCIAL AND BEHAVIOURAL SCIENCES

Managing Editor:

JAAKKO HINTIKKA, *Academy of Finland and Stanford University*

Editors:

ROBERT S. COHEN, *Boston University*

DONALD DAVIDSON, *University of Chicago*

GABRIËL NUCHELMANS, *University of Leyden*

WESLEY C. SALMON, *University of Arizona*

VOLUME 115

ROGER D. ROSENKRANTZ

Virginia Polytechnic Institute and State University, Blacksburg, U.S.A.

INFERENCE, METHOD AND DECISION

Towards a Bayesian Philosophy of Science

D. REIDEL PUBLISHING COMPANY

DORDRECHT-HOLLAND / BOSTON-U.S.A.

Library of Congress Cataloging in Publication Data

Rosenkrantz, Roger D.   1938–
    Inference, method, and decision.

    (Synthese library; v. 115)
    Includes bibliographies and indexes.
    1.   Science—Philosophy.   2.   Bayesian   statistical   decision
theory.   I.   Title.
Q175.R554   501   77–9399
ISBN 90–277–0817–7
ISBN 90–277–0818–5 pbk.

Published by D. Reidel Publishing Company,
P.O. Box 17, Dordrecht, Holland

Sold and distributed in the U.S.A., Canada, and Mexico
by D. Reidel Publishing Company, Inc.
Lincoln Building, 160 Old Derby Street, Hingham,
Mass. 02043, U.S.A.

Printed in The Netherlands

To Margaret

PREFACE

This book grew out of previously published papers of mine composed over a period of years; they have been reworked (sometimes beyond recognition) so as to form a reasonably coherent whole.

Part One treats of informative inference. I argue (Chapter 2) that the traditional principle of induction in its clearest formulation (that laws are confirmed by their positive cases) is clearly false. Other formulations in terms of the 'uniformity of nature' or the 'resemblance of the future to the past' seem to me hopelessly unclear. From a Bayesian point of view, 'learning from experience' goes by conditionalization (Bayes' rule). The traditional stumbling block for Bayesians has been to find objective probability inputs to conditionalize upon. Subjective Bayesians allow any probability inputs that do not violate the usual axioms of probability. Many subjectivists grant that this liberality seems prodigal but own themselves unable to think of additional constraints that might plausibly be imposed. To be sure, if we could agree on the correct probabilistic representation of 'ignorance' (or absence of pertinent data), then all probabilities obtained by applying Bayes' rule to an 'informationless' prior would be objective. But familiar contradictions, like the Bertrand paradox, are thought to vitiate all attempts to objectify 'ignorance'. Building on the earlier work of Sir Harold Jeffreys, E. T. Jaynes, and the more recent work of G. E. P. Box and G. E. Tiao, I have elected to bite this bullet. In Chapter 3, I develop and defend an objectivist Bayesian approach.

Given a number of constraints, say, mean values or symmetries, Jaynes' *maximum entropy rule* singles out that prior distribution which, among all those consistent with the given constraints, maximizes entropy (a measure of uncertainty). Far from 'generating knowledge out of ignorance', this rule enjoins us not to pretend to knowledge we do not possess. I regard it as a postulate of rationality, on a par with the 'sure-thing principle' or the expected utility rule.

Jaynes shows that, of all distributions consonant with the given constraints, the maximum entropy distribution is realized in by far the highest proportion of possible experimental outcomes. Consequently, if an actual outcome fails to fit the maximum entropy distribution at all well, one can be practically certain either that additional constraints are operative or that the given constraints are not wholly correct. (In that event, the initial distribution is not revised by conditionalization, but retracted.) To many of my readers, this will have a whiff of rationalism about it, but I think it a defensible rationalism. The given constraints are presumably empirical, but given their correctness and exhaustiveness, the correct distribution should be discoverable by pure thought in the same way that abstract argument leads one to the normal distribution for a trait that depends on a large number of independent factors. I contend there are no conceptual distinctions to be drawn between data distributions and prior distributions; both are arrived at by the same kind of reasoning and both are equally 'subjective' or 'objective'. Whether a given distribution counts as one or the other is largely a function of the use to be made of it. For this reason, maximum entropy distributions are often amenable to empirical check, and where they fail, other or different constraints are invariably discovered. Further justification of Jaynes' rule is found in its essential agreement with another rule of Sir Harold Jeffreys', and a more general rule recently devised by Box and Tiao. The confluence of these methods (which are based on quite disparate intuitions) is evidence of their soundness in the same way that the agreement of different definitions of 'recursive' evidences Church's thesis.

Invariance requirements are especially powerful constraints, often sufficient to uniquely determine a distribution. Jaynes has generated a prior for the Bertrand problem by an invariance argument and empirically verified it. In actual fact, absence of data rarely, if ever, connotes total ignorance, and consequently, it is a mistake to suppose that any old transformation of parameters is admissible. We must rather uncover the admissible re-parametrizations by asking for the equivalent reformulations of a problem, and these will often be implicit in the statement of the problem itself. Thus, we might be asked for the probability that k is the first significant digit of an entry in a table of data (say, the areas of the world's largest islands), $k = 1, \ldots, 9$. Since nothing is said about the scale units employed, scale invariance is implied. That is, if there were such a 'law of first digits', it should be independent of scale.

Given the intuitive adequacy of the rules for generating 'no data' priors, the objectivist position begins to seem quite tenable. Even conditionalization has a derivative status, from this perspective. Any correct rule for revising probabilities in the light of the additional information that the 'conditioning event' has occurred should have the property of leading from a correct representation of the one state of partial knowledge to a correct representation of the other. That Bayes' rule does have this property is compelling evidence that it is the (uniquely) correct solution of the 'kinematical' problem.

From the standpoint of objectivist Bayesianism, one compares theories by their (objective) probabilities. I argue, in Part Three, that probabilities are sufficient to guide research, and against the advisability of imposing formal rules of acceptance and rejection for theories. It is also a corollary of a thoroughgoing Bayesian philosophy of science that other traits of theories which have been thought important enter objectively only to the extent that they are reflected in higher probability. Simplicity (measured in a way that generalizes the familiar 'paucity-of-parameters' criterion) is shown to be such a trait in Chapter 5, and the implications of this fundamental observation are traced out in the remainder of Part Two (and are felt in Part Three, as well). Since the detailed discussion of these matters is rather technical, a brief and more intuitive account may be useful to the reader.

I show, specifically, that in a broad spectrum of cases, the 'simpler' of two theories in equally good agreement with the data is better supported, and so also, confirmed. 'Support' and 'confirmation' are used throughout in a precise sense related to probability. A theory is *confirmed* by an observation when its probability is increased thereby. As for 'support', suppose that we have a partition of hypotheses, H_i (exactly one must be true). Then the relative support accorded them by an observation x is registered by the conditional probabilities, $P(x/H_i)$. Considered as a function of an hypothesis H, $P(x/H)$ is called the *likelihood function* (associated with x). The 'likeliest' or best hypothesis maximizes the likelihood function; it accords what was observed the highest probability. Bayes' rule for revising probabilities states that the new (or 'posterior') probability of H is proportional to the prior probability, $P(H)$, and its likelihood, $P(x/H)$, and inversely proportional to $P(x)$, in symbols: $P(H/x) = P(H)P(x/H)/P(x)$.

Likelihood, then, measures the impact of the data on probability: the posterior probability is a blend of the initially given information, as represented by $P(H)$, and the sample information, as embodied in the likelihood function. Of two competing hypotheses, H and K, the 'likelier' will be confirmed (its probability will go up, while that of the other hypothesis goes down). Indeed, using Bayes' rule, $P(H/x) : P(K/x) = [P(H) : P(K)] [P(x/H) : P(x/K)]$, since $P(x)$ cancels. This expresses the *posterior odds* as a product of the *prior odds*, $P(H) : P(K)$, and the *likelihood ratio*, $P(x/H) : P(x/K)$.

Scientific theories often contain adjustable parameters which must be estimated from the data each time the theory is applied. How might the support of two such theories be compared? One approach, often used, is to compare the support of their 'likeliest' or 'best-fitting' special cases (obtained by fitting the free parameters). The two theories, however, may differ in the number of their free parameters, and typically, a theory with more free parameters will fit more possible observations (it will be less 'simple', in my sense). This feature is surely relevant to any comparative evaluation of the theories. But it is unclear, at a purely judgmental level, how much to discount for a theory's comparative lack of simplicity when comparing two theories in the indicated way.

A Bayesian solution of the problem would take the following form. Let the two hypotheses, H and K, have special cases H_1, H_2, H_3 and K_1, K_2, K_3 with prior probabilities and likelihoods as shown below:

$$P(x/H_1) = 1/5, P(x/H_2) = 1/10, P(x/H_3) = 1/20 ,$$

$$P(H_1) = 6/20, P(H_2) = 3/20, P(H_3) = 1/20 ;$$

$$P(x/K_1) = 1/4, P(x/K_2) = 1/6, P(x/K_3) = 1/12 ,$$

$$P(K_1) = 3/20, P(K_2) = 4/20, P(K_3) = 3/20 .$$

Here I am assuming, of course, that the hypotheses $H_1 - H_3$, $K_1 - K_3$ are mutually exclusive and jointly exhaustive. Since a theory holds if some special case of it holds, the probability of a theory is the sum of the probabilities of its special cases (or, more accurately, of its 'ultimate' special cases, obtained by fixing the values of all its free parameters). To obtain posterior probabilities, we must therefore sum the posterior probabilities of the theory's special cases. Since $P(H_i/x) \propto P(x/H_i)P(H_i)$, we have:

$$P(H/x) \propto P(x/H_1)P(H_1) + P(x/H_2)P(H_2) + P(x/H_3)P(H_3)$$
$$+ (1/5) (6/20) + (1/10) (3/20) + (1/20) (1/20) = 0.0775.$$

Likewise, $P(K/x) \propto 0.0833$. Since $P(H/x$ and $P(K/x)$ must continue to sum to 1, $P(H/x) = 0.4819$ and $P(K/x) = 0.5181$. Hence, K is better supported (and so confirmed), its probability increasing ever so slightly, while that of H suffers a corresponding decrement.

We have obtained the support by averaging over the likelihoods of the special cases of the theory, each such likelihood being weighted by the probability of the corresponding special case. When we add a parameter, complicating a theory, we add to the number of special cases over which we must average the likelihood function, and because comparatively few of its special cases will fit the data at all well, the average likelihood is degraded.

Both accuracy and simplicity are therefore reflected in a theory's average likelihood or support, but in a mathematically determinate way that obviates the need for a judgmental assessment of the relative importance of these two factors.

Given the determinate rate of exchange between accuracy and simplicity, we can say by how much a theory's accuracy must be improved in complicating it for its support to increase. (That the support increases means that the accuracy gained is sufficient to offset the simplicity lost.) For example, a circular orbit may fail to fit the plotted positions of a planet very well, and the question arises whether an elliptical orbit (which contains the eccentricity as an additional parameter) would be better supported. (To compare the average likelihoods in this case requires a probabilistic treatment of the observational errors.) A circle is, of course, a special case of an ellipse, and so the two hypotheses are not logically exclusive. Strictly speaking, Bayes' rule can only be applied to a comparison of exclusive alternatives, but the difficulty can be circumvented in practice by taking logical differences: e.g., we compare the hypothesis of a circle to that of a *proper* ellipse (viz. an ellipse of positive eccentricity). The method of average likelihood is applied to some interesting examples from genetics in Chapter 5 and to curve-fitting and related problems in Chapter 11, where its performance is compared with that of several non-Bayesian methods.

One additional point is worth noting. It is evident that the average of the likelihood function can never exceed its maximum value. This observation is mathematically trivial, yet fraught with methodological significance. For it translates into the statement that a theory (with parameters) can never be better supported than its best-fitting (or 'likeliest') special case. We will see (Chapter 7) that the two main arguments Copernicus marshalls on behalf of the heliocentric hypothesis are but applications of this principle. And the Copernican example illustrates the applicability of Bayesian methods to deterministic, as well as to probabilistic, theories, for the former typically include a probabilistic treatment of error.

One friend who read this book in manuscript remarked that it seemed to fall between two stools, for many of the scientific examples are akin to those one might encounter in a statistics text, while others are more like those found in works on the history and philosophy of science. This aspect of the book is not unintentional. My deliberate aim here is to bring these formerly disparate areas of study together in one unified treatment – whence my subtitle. The symbiosis should be mutually enriching, and of interest to philosophers and historians of science, scientists, and statisticians alike.

The Bayesian position which I develop and the applications of it to issues in the philosophy of science owes much to the writings of I. J. Good and E. T. Jaynes. If the book makes their important work accessible to a wider philosophical public, I will not have labored in vain. To Good, I am also indebted for much stimulating conversation and many useful comments.

My early interest in the subject developed while I was a graduate student at Stanford, and the members of my dissertation committee, Professors Patrick Suppes, Jaakko Hintikka and Joseph Sneed, provided an especially exciting and supportive environment in which to develop my ideas. They have continued to provide much stimulation and support in the intervening years. My thinking was further shaped by participation in a discussion group organized by the late Allan Birnbaum. Allan encouraged my own tendency to view statistics and philosophy of science as a connected whole, and, more than that, to test one's views and methods in the crucible of actual scientific research. His patient but sharp criticisms of Bayesian methods incited me to investigate their applicability in a serious way and the fruits of that search are scattered through Chapters 5, 7, 9, 10 and 11. The other members of the group, Ronald Giere and Isaac Levi, have also done much to help me avert minor skirmishes and focus critical attention on the more significant issues that separate my position from theirs.

The list of those who have influenced me could go on and on, but I would be remiss indeed if I failed to mention Stephen Spielman, Barry Loewer, W. K. Goosens, Kenneth Friedman, and my former student, Robert Laddaga. Their characteristically insightful remarks have helped to advance my thinking at crucial points, forced its clearer expression, and saved me from serious errors. I owe them all a very large debt.

Earlier versions of Chapters 5, 7, and 9 were delivered at the 1973 and 1975 University of Western Ontario Workshops organized by W. A. Harper, C. K. Hooker and James Leach. A talk that evolved into Chapter 1 was delivered at Ernest Adams' Berkeley seminar, while another which grew into Chapter 11 was delivered at Patrick Suppes' probability seminar at Stanford (both in 1975). I am grateful to all of these men for the opportunity to try out my ideas, and to the participants in those seminars and workshops for their many useful comments and criticisms.

I would also like to thank Mrs. Dabney Whipple for her patient and accurate typing of the manuscript and her help in preparing it for publication. Last, but far from least, my wife, Margaret, to whom the book is dedicated, programmed the empirical Bayes/orthodox comparisons of Chapter 11 and provided the moral and material support without which the book could not have been written.

TABLE OF CONTENTS

PART ONE

INFORMATIVE INFERENCE

CHAPTER 1

INFORMATION

1. INTRODUCTION

We ordinarily think of induction as 'learning from experience', and the question naturally arises how to make such learning proceed most rapidly. The answer, in rough and ready terms, is to ask searching questions, and experimentation is conceived, in this spirit, as the art of asking such questions or probing Nature for her secrets. Now searching questions are those which promise to shed most light on the problem of interest, in a word, to deliver the highest expected yield of information. This already suggests the relevance of information theory, but it does not establish it, for information theory was developed by communication theorists and engineers to solve problems whose connection with efficient experimentation is less than obvious. Yet, as we will see, the connections are there all right, and it is part of our task in this chapter to articulate them.

Information theory bears perhaps most directly on the problem of efficient coding. It is tempting to construe theories as more or less efficient ways of encoding data. This direction of inquiry will be pursued in Chapter 8, but it is worth remarking here that efficient coding and efficient experimentation are formally identical problems. That is, maximizing the average information transmitted per message sent is formally identical to maximizing the average information per experimental trial about the true state of the world (or transmitted message).

Research in a mature scientific discipline is no mere random groping or haphazard accumulation of facts, but a highly directed activity. Our theories tell us (roughly) what the world is like, what to look for, and sometimes even where to look — witness the discovery of Neptune or of the 'missing' elements of the Periodic Table. The workaday projects that fill the lives of most scientists are aimed, as Kuhn has rightly stressed, at filling out a theoretical blueprint, or at supplying additional evidence for a theory, using fairly standard experimental and mathematical techniques. Calculating perturbations or plotting an orbit, mapping a chromosome, or measuring a physical constant, are all problems of this sort. In any event, the theories that interest us determine what facts will interest us, and pursuing this line of thought leads one naturally to treat experimentation as a quest for *useful* or *salient* information, and hence, as an essentially decision theoretic problem.

I have already stated my opposition to attempts to reduce theory appraisal to rational decision making, and that opposition carries over to attempts to treat the problem of efficient experimentation in decision theoretic terms. That opposition, I hasten to add, is qualified: where practical concerns enter, decision theory is, of course, appropriate. And I would also admit that practical concerns typically do enter scientific work, if only in the form of sampling costs. But, in pure research, even where experimental costs enter, the utilities assigned different experiments should be monotonic functions of the expected disinterested information yielded by those experiments. In short where pure research is in question, the aim of finding out as much as possible without regard to whether the information obtained is good news must be reflected in our utility assignments. We are led, then, to develop a theory of disinterested information. Those who maintain that all research is ultimately directed towards practical concerns can regard information as 'disinterested' when it bears, in foreseen or unforeseen ways, on a variety of such concerns or practical decisions.

Although my interest in decision theoretic treatments of experimentation is peripheral, the subject does have a certain importance in its own right (whatever its bearing on the correct understanding of the scientific enterprise). It will be worthwhile to see what relation, if any, this theory has to the various measures of disinterested information we shall be studying. Finally, some familiarity with decision theory is prerequisite to Chapters 10 and 12, where assorted decision theoretic conceptions of scientific activity are subjected to critical scrutiny.

2. THE UTILITY OF INFORMATION

This section concerns *risky decision making*, or decision making with incomplete knowledge of the state of nature. For example, the decision might involve classifying a patient as infected or uninfected, marketing or withholding a new drug, or determining an optimal allocation of stock. When one of the options is best under *every* possible circumstance, the choice is clear (the so-called 'sure-thing principle'). In general, though, the best course of action depends on which state of nature obtains. It is clear that if one has fairly sharply defined probabilities for the different states, and fairly well defined views on the desirability of performing the several actions under the considered states, then the best action is that which has highest utility at the most probable states. If numerical utilities and probabilities are assigned, we are led to a sharper, quantitative form of this principle; *choose that action which maximizes expected utility*. The *expected utility* of an action is the weighted average of its utilities under the several states, the weights being the respective probabilities of those states. The rule in question is variously referred to as the *expected utility rule* or the *Bayes decision rule*. An action which is best by the lights of the rule (i.e., an action which maximizes expected utility) is called a *Bayes act*.

In many cases, numerical utilities can be identified with monetary payoffs for practical purposes. But utility cannot generally be identified with money; it depends on such additional factors as the prospective uses to which the money is put, levels of aspiration, externalities, risk averseness, and, of course, on the agent's initial fortune. Thus, ten dollars generally has more utility for a pauper than for a millionaire, and more utility still for a pauper who needs just ten dollars more to realize a life-long ambition. Moreover, it may not be easy to find a monetary equivalent for the dire consequences of an inappropriate decision (which might even result in death). We shall not enter into these and other complications here, since our interest is largely confined to the bearing of probability and probability changes on decisions taken. In what follows, therefore, I take utilities (or payoffs) as given, but my treatment of probabilities will be more realistic. To see right off how the Bayes rule may apply where probabilities are incompletely known, consider a simple two-act two-state decision problem, whether or not to invest $5000.00 in a corporate stock, with payoffs given in Table I.

TABLE I

	Conditions favorable	Conditions unfavorable
Invest	20 000	− 5000
Don't invest	0	0

The expected value of investment when p is the probability that business conditions are favorable is $20\,000p - 5000(1 - p)$, which is positive if $p > 1/5$. Since the expected value of the other option is zero, it pays to invest if $p > 1/5$. Hence, you do not need to attach a precise numerical value to the probability p in order to reach a decision. The value $1/5$ is called the *critical probability*; it is the break-even point.

To fill out the account, we must admit the possibility of making additional observations or collecting more information. Our aim here is to measure the utility of information from an experiment. Obviously, the utility of an experiment is the average of the utilities of its different possible outcomes, the weights being the probabilities of those outcomes. So the problem reduces to measuring the utility of a particular outcome.

Now each outcome x will lead to a new (posterior) probability distribution P_x of the states. An action which maximizes expected utility against P_x is called a *posterior Bayes act associated with* x and designated a_x. The act which is Bayes against the prior distribution – the *prior Bayes act* – will be designated $\mathbf{a}^*$. If $a_x = \mathbf{a}^*$, I call x *ineffective*; otherwise x is called *effective*. An ineffective outcome is one which does not change the agent's initial choice of action, and hence it has no cash value or utility for the decision problem at hand. (Of course, an ineffective outcome might be effective for some other decision problem; the notion of effectiveness is relativized to a decision problem.) It remains only to measure the utility of an effective outcome.

Once we observe the outcome of the experiment, we find ourselves in a new state of (partial) belief reflected by the new distribution P_x. The utilities of act-state pairs remain the same, however, and so we measure the utility of x by the expected utility of doing a_x instead of $\mathbf{a}^*$, given P_x. Writing U^* for the expected utility, and $U^*(a/P_x)$ for the expected utility of a against the posterior distribution P_x, the utility of x, called the *conditional value of sample information* (or *CVSI*) associated with x, is measured by:

(1.1) $CVSI(x) = U^*(a_x/P_x) - U^*(a^*/P_x)$.

Since a_x is a Bayes act for outcome x (i.e., maximizes $U^*(a/P_x)$), the *CVSI* is non-negative. More precisely, it is positive for effective outcomes and zero for ineffective outcomes. The value of the experiment e of which x is a typical outcome, called the *expected value of sample information* from e (or *EVSI*) is the outcome-probability weighted average of the *CVSI*'s of the several outcomes:

(1.2) $EVSI(e) = \Sigma_x CVSI(x) P(x)$.

As a weighted average of non-negative quantities, the *EVSI* is always non-negative. Let e^* be a *perfect experiment* (each outcome of e definitively identifies the true state) and let e_0 be the *null experiment*. The latter can be characterized either as a no-experiment or an experiment some particular one of whose outcomes is certain to occur. Write *EVPI* (*expected value of perfect*

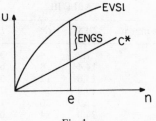

Fig. 1.

information) and *EVNI* (*expected value of no information*) for $EVSI(e^*)$ and $EVSI(e_0)$, respectively. Then it is clear that for any experiment e we have:

(1.3) $0 = EVNI \leqslant EVSI(e) \leqslant EVPI$.

Let $C^*(e)$ be the expected cost [1] of e (e might be a sequential experiment with a probabilistic stopping rule.) Then, the *expected net gain of sampling* (or *ENGS*) is defined by:

(1.4) $ENGS(e) = EVSI(e) - C^*(e)$.

The optimal experiment for a given decision problem maximizes *ENGS*. The case of fixed sample size experiments with constant cost per item sampled is shown in Figure 1, costs and payoffs being measured, of course, on the same utility scale. The optimal e is shown.

EXAMPLE 1 (Raiffa (1968)). Type 1 urns contain 4 red and 6 black balls; type 2 urns contain 9 red and 1 black. A given urn is drawn from a collection of which 80% are type 1 and 20% are type 2. Let s_i be the state that the chosen urn is type i and let a_i be the act of guessing type i, $i = 1,2$. The monetary payoffs are shown in Table II. Here $a^* = a_1$ with $U^*(a^*) = 28$. More generally, if $p = P(s_1)$ is unknown, $U^*(a_1) = 60p - 20$ and $U^*(a_2) = -105p + 100$, and the critical probability is 8/11 (i.e., a_1 is the prior Bayes act if p exceeds 8/11).

TABLE II

	s_1	s_2
a_1	40	-20
a_2	-5	100

TABLE III

Experiment	EVSI	C*	ENGS
e1	7.2	4	3.2
e2	14.4	8	6.4
e3	15.6	12	3.6
e4	19.7	16	3.7
e5	21.9	20	1.9
e6	24.0	24	0.0

Now you are given the option of sampling balls at random (without replacement) from the chosen urn at \$4 per ball. It is found that e_2 (sampling two balls) is optimal (see Table III). $EVSI$ increases here at a generally diminishing rate, while C^* increases linearly, the largest gap between them coming at e_2 (which therefore maximizes $ENGS$). I.e., sampling just two balls provides the most valuable information for this problem per unit cost. We also have $e^* = e_6$: by the time you have sampled 6 balls, whatever the outcome, you *know* whether the urn is type 1 or type 2. $C^*(e^*) = EVSI(e^*)$, but, more commonly, the perfect experiment (which may entail sampling the entire population) has negative $ENGS$. We illustrate the calculations, using e_2.

Only the outcome RR (both balls red) is effective, since all of BB, BR and RB are more likely to come from a type 1 urn, confirming us in our original choice of a_1. The relevant likelihoods are: $P(RR/s_1) = 6/45$, and $P(RR/s_2) = 36/45$, whence $P(RR) = P(RR/s_1)P(s_1) + P(RR/s_2)P(s_2) = 12/45$. By Bayes' theorem, $P_{RR} = (0.4, 0.6)$ gives the posterior probabilities of s_1 and s_2. From the latter we find $CVSI(RR) = U^*(a_2/P_{RR}) = 0.4(-5) + 0.6(100) - (0.4(40 + 0.6(-20)) = 54$. Hence, $EVSI(e_2) = 54 \times 12/45 = 14.4$. A sequential experiment for this problem is discussed and its $ENGS$ calculated in Raiffa (1968), Chapter 2.

Write u_{ij} for $U(a_i, s_j)$, $p_i = P(s_i)$, and $U_k = U^*(a_k)$. (We are dealing now with the general case.) We have:

$$(1.5) \qquad U_h - U_k = \Sigma_i (u_{hi} - u_{ki}) p_i \geqslant (u_{hj} - u_{kj}) p_j$$
$$+ (1 - p_j) \min_{i \neq j} (u_{hi} - u_{ki}).$$

Suppose first that a_h is best at s_j, so that the first term on the right is positive. Then $U_h - U_k \geqslant 0$ if $p_j \geqslant -\min_{i \neq j}(u_{hi} - u_{ki})/(u_{hj} - u_{kj} - \min_{i \neq j}(u_{hi} - u_{ki}))$. Hence a_h is Bayes if p_j exceeds the largest of these terms. Suppose next that a_h is best at every state save s_j, so that the first term on the right side of (1.5) is negative for some k and the second term is positive. Then $U_h - U_k \geqslant 0$ if

$(1 - p_j)\min_{i \neq j}(u_{hi} - u_{ki}) \geqslant -p_j(u_{hj} - u_{kj})$, or $p_j \leqslant -\min_{i \neq j}(u_{hi} - u_{ki})/(u_{hj} - u_{kj} - \min_{i \neq j}(u_{hi} - u_{ki})$. Hence, a_h is Bayes in this case if p_j does not exceed the smallest of these terms. We have proved.

THE THEOREM ON CRITICAL PROBABILITIES (Bragga-Illa). Given

$$(1.6) \qquad -\min_{i \neq j}(u_{hi} - u_{ki})/(u_{hj} - u_{kj} - \min_{i \neq j}(u_{hi} - u_{ki})) \,,$$

write $p_j{}^{**}$ for the maximum of the terms (1.6) for $k \neq h$ and p_j* for the minimum of these terms for $k \neq h$. Then:

(i)   If a_h is best at s_j, a_h is Bayes if $p_j \geqslant p_j{}^{**}$.
(ii)   If a_h is best at every state save s_j, a_h is Bayes if $p_j \leqslant p_j{}^*$. $p_j{}^{**}$ and $p_j{}^*$ are called *upper* and *lower critical probabilities*, resp.

EXAMPLE 2. The utility matrix is:

TABLE IV

	s_1	s_2	s_3
a_1	10	−6	0
a_2	−5	15	−7
a_3	−8	−10	20

Here a_h is best at s_h, $h = 1, 2, 3$, and I calculate $p_1{}^{**} = 7/12$, $p_2{}^{**} = 27/52$, and $p_3{}^{**} = 25/52$. Note, $p_h{}^{**}$ decreases with the amount by which a_h beats its rivals at s_h and increases with the maximum amount by which a_h is beaten at the remaining states. The critical probabilities embody all the relevant comparisons in a compact form.

We usually attach a positive utility to the receipt of information *per se*, and *a fortiori*, to the receipt of good news. By the same token we attach disutility to the receipt of bad news; yet no conflict with the non-negativity of the *EVSI* is implied. A student might prefer not to know his final grade, a man who is seriously ill might prefer not to know exactly how long he has. Similarly, a man might rather not know what another is saying behind his back. But the expected utility of information can only be assessed where specific practical decisions are contemplated. Thus, the interest of the last man would increase markedly were he contemplating whether to ask the other for a letter of recommendation! The point is that, while disutility would attach to the information that he was saying unkind things, ever so much greater disutility would attach to his saying those unflattering things in a letter of recommendation. Consequently, net expected utility will attach to finding out what he is already saying to others, as shedding light on what he is likely to write about one. The next example (from Mosteller, 1965) sheds additional light on the matter.

EXAMPLE 3. In this problem, there are three prisoners A,B,C two of whom are to be released. Prisoner A calculates his a priori chances at 2/3. The warder agrees to tell him the name of one prisoner other than himself who is to be released. Write b,c for the two possible answers. If b (the warder says B) one of the three states (A and C released) is excluded, and so A re-calculates his chances at 1/2 in the light of b. The same chances are obtained for c, and for the same reason. Hence, on either outcome of his little experiment, A finds that his probability of release has declined and accordingly concludes that the experiment has negative expected utility. Where has A erred in his calculations? (I leave the solution to you.)

In other cases, it is seemingly rational to refuse information on the grounds that it is likely to be misinformation. Studies of comparative intelligence, especially those comparing the intelligence of different races or ethnic groups, are discouraged on such grounds. Doubtless, cases of this sort pose delicate and sensitive issues for public policy, but my interest here is confined to the question whether they are in any way incompatible with the non-negativity of the *EVSI*.

Conflict can only arise in connection with a practical decision problem. Does our mathematics force the conclusion that even a highly misleading I.Q. test is better than none in deciding what teaching methods to employ?

EXAMPLE 4 (an unreliable test.) Students are classified as either bright or dull using an I.Q. test, and different teaching methods, t_1 and t_2, are applied to the two groups. Write a_i for the decision to classify a student as type i, and let s_i be the student's actual state of intelligence, $i = 1$ (bright), 2 (dull). The utility matrix is:

TABLE V

	s_1	s_2
a_1	10	−20
a_2	−20	10

Let p be the proportion (partially known) of bright students. To simplify the calculations, I will assume that the test has equal probabilities of misclassifying a bright student as dull or a dull student as bright. Write α for this error probability. Then the expected utility of using the test (where the status quo — teaching all students as before — is assigned utility 0) is given by:

$$p(1 - \alpha)U(a_1, s_1) + (1 - p)\alpha U(a_1, s_1) + p\alpha\, U(a_2, s_1)$$
$$+ (1 - p)(1 - \alpha)(Ua_2, s_2)$$

where $p(1 - \alpha)$ is the proportion of students the test correctly classifies as bright, etc. If $p = 0.5$, this quantity reduces to $10\,(1 - \alpha) - 20\alpha$, which is negative if α exceeds 1/3 (the break-even point). That is, the status quo becomes preferable to using the test (and more specialized teaching) when the probability of misclassifying a student exceeds 33%. A test which is likely to have that high an error rate should not be used. Far from showing that a misleading I.Q. test is better than none, our analysis shows the opposite.

3. DISINTERESTED INFORMATION

We take up now the theory of disinterested information; the problem here is to maximize information without regard to its utility for any particular decision problem. We base our treatment on the obvious analogy between experimentation and communication over a noisy channel. The parameter space of an assumed model plays the role of source or message ensemble from which the input, the state of nature or setting of the parameters, is selected. The input is then transmitted by the experiment to the target, the experimenter, who proceeds to decode the message. Noise enters in the form of sampling error, systematic bias, the masking effects of hidden variables, uncontrolled variation in the experimental materials, and so forth. The information transmitted from source to target measures the average amount by which the experiment reduces the experimenter's uncertainty regarding the parameters of his model. Thus, for a fixed model, each of the available experiments has an associated expected yield of information. If, as we assume throughout, information is the goal of research, then the experimenter should, by performing the experiment with the highest expected yield of information, select the least noisy of the available channels. In a straightforward sense, that experiment can be regarded as most sensitive, or as providing the weightiest evidence.

Transmitted information does not depend on the direction of flow, that is, on whether the parameter space of the model or the outcome space of the experiment plays the role of source. The symmetry suggests that, for purposes of predicting the outcome of a fixed experiment, that model should be preferred which transmits, on the average, a maximal quantity of information regarding the outcome space. Such models can be said to be *maximally informative* with respect to the experiment.

The first thing we need in carrying out the projected development is a suitable measure of uncertainty. Let X be a discrete[2] random variable with possible values x_i which have probabilities p_i, $i = 1, \ldots, m$. By the *entropy* of X (or its distribution) is meant the quantity:

$$(1.7) \qquad H(p_1, \ldots, p_m) = -\Sigma_i p_i \log p_i \, .$$

Logarithms may be taken to any base $b > 1$. While we confine discussion to the discrete case, theorems given here can be extended to continuous random variables.[3]

When logarithms are taken to the base 2, the unit of the entropy scale is called a 'bit' – short for 'binary digit'. A one bit reduction in H is brought about by eliminating half of an even number of equiprobable alternatives.

A more illuminating motivation for (1.7) can be given by noting that $H = log \, m$ *is the minimum number of binary digits needed to code answers to* m *yes/no questions*. E.g., to locate a single square on a checkerboard (containing 64 squares), a minimum number of $6 = \log_2 64$ questions is required, and the minimum is attained just in case each question splits the remaining alternatives in half (or, equivalently, maximizes the uncertainty of the response). Thus, the first question reduces uncertainty to 32 alternatives, the second to 16 alternatives, and so on. And the answers to each of these questions have equal probabilities of being given, a priori. A binary sequence of six digits, like '001100', could be used to code the answers, so that each sequence would correspond biuniquely to a square on the board (which you can think of as an unknown state of nature or transmitted message).

Given m equiprobable alternatives, each has probability $p = 1/m$; for this case, $H = -\Sigma p \log p = -mp \log p = -\log p$. In other words, the average uncertainty associated with each alternative is $h_i = -\log p$. Consider next the case of unequally probable alternatives. For example, let E_i be the event that a pair of ordinary dice turn up a sum of i spots, $i = 2,3, \ldots, 12$. Suppose we restrict the outcome space to the event E_7 ('sevens'). Then our uncertainty is reduced to six equiprobable alternatives – the ordered pairs (1,6), (6,1), (2,5), (5,2), (3,4), (4,3) – and is measured by $h_7 = \log 6 = -\log P(E_7)$. And, in general, when restricted to E_i, our uncertainty is measured by $-\log P(E_i)$. Consequently, our uncertainty regarding the partition

$E_2, \cdots, E_{12}$ is given by the weighted average of the h_i, or:

$$H = \Sigma_i h_i P(E_i) = -\Sigma_i p_i \log p_i,$$

that is, by (1.7).

For our example, $p_i = (i-1)/36$ for $i = 2, \ldots, 7$ and $p_i = (12 - (i-1))/36$ for $i = 8, \ldots, 12$. Hence $H = 3.71$ bits. The reader can verify that a minimum number of three or four yes/no questions is required to ascertain which of the alternatives $E_2, \ldots, E_{12}$ obtains.

That entropy is a satisfactory measure of uncertainty is further attested by the following of its properties:

(i) $H(p_1, \ldots, p_m) = H(p_1, \ldots, p_m, 0)$, the entropy is wholly determined by the alternatives which are assigned a non-zero probability.

(ii)   When all the p_i are equal, $H(p_1, \ldots, p_m)$ is increasing in m, the number of equiprobable alternatives.

(iii) $H(p_1, \ldots, p_m) = 0$, a minimum, when some $p_i = 1$.

(iv) $H(p_1, \ldots, p_m) = \log m$, a maximum, when each $p_i = 1/m$.

(v)   Any averaging of the p_i (i.e., any flattening of the distribution) increases H.

(vi) H is non-negative.

(vii) $H(p_1 \ldots, p_m)$ is invariant under every permutation of the indices $1, \ldots, m$.

(viii) $H(p_1, \ldots, p_m)$ is continuous in its arguments.

The informational measures that interest us presuppose the concepts of joint entropy and conditional entropy. The *joint entropy* of X and Y is defined by:

$$(1.8) \qquad H(X,Y) = -\Sigma_i^m \Sigma_j^n p_{ij} \log p_{ij}, \qquad p_{ij} = P(X = x_i, Y = y_j)$$

and measures the joint uncertainty (or informational overlap) of X and Y. The *entropy* $H_i(Y) = H_{x_i}(Y)$ *of Y conditional on* x_i is given by:

$$(1.9) \qquad H_i(Y) = -\Sigma_j p_i(j) \log p_i(j), \qquad p_i(j) = P(Y = y_j / X = x_i)$$

and measures the uncertainty in one random variable, Y, when the value of another random variable, X, is known. $H_i(Y)$ is itself a random variable defined on the range of X, and its expected value

$$(1.10) \qquad H_X(Y) = \Sigma_i p_i H_i(Y) = -\Sigma_i p_i \Sigma_j p_i(j) \log p_i(j)$$

is called the *entropy of Y conditional on X*, and measures, of course, the average uncertainty in Y given that some one of the states $X = x_i$ obtains.

Since $\Sigma_j p_i(j) = 1$,

$$-H(X, Y) = \Sigma_{ij} p_{ij} \log p_{ij} \,,$$

$$= \Sigma_i \Sigma_j p_i(j) \log p_i(j) + \Sigma_j p_i(j) \, \Sigma_i p_i \log p_i \,,$$

$$= -H_X(Y) - H(X) \,,$$

writing $p_{ij} = p_i(j) p_i$, whence

(1.11) $H(X,Y) = H_X(Y) + H(X) \,,$

which is to say the uncertainty in both X and Y is obtained by adding the uncertainty in X to the uncertainty in Y given X. The identity (1.11) is the additive analogue of the multiplicative relation $P(A,B) = P(B/A)P(A)$ utilized in its proof. By writing the joint probability p_{ij} in the alternative form $p_j(i)q_j$, where $q_j = P(Y = y_j)$, we obtain

(1.12) $H(X,Y) = H_Y(Y) + H(Y)$

as an alternative form of (1.11). Together (1.11) and (1.12) express the symmetry of joint entropy: $H(X,Y) = H(Y,X)$.

Exploiting the convexity of $g(t) = t \log t$, i.e., that $\Sigma_i p_i g(t_i) \geqslant g(\Sigma_i p_i t_i)$, by the substitution $t_i = p_i(j)$, we get

$$\Sigma_i p_i p_i(j) \log p_i(j) \geqslant \Sigma_i p_i p_i(j) \log (\Sigma_i p_i p_i(j)) = q_j \log q_j \,,$$

whence

$$\Sigma_i p_i \Sigma_j p_i(j) \log p_i(j) \geqslant \Sigma_j q_j \log q_j \,,$$

or

$$-H_X(Y) \geqslant -H(Y) \,,$$

which is equivalent to *Shannon's inequality:*

(1.13) $H_X(Y) \leqslant H(Y)\,.$

From the proof it is obvious that equality holds just in case X and Y are independent. Loosely expressed, (1.13) asserts that uncertainty never increases (on the average) by additional specification of the state of nature.

We can now round out our list of the fundamental properties of entropy with the following:

(ix) $H(X,Y) = H_X(Y) + H(X) = H_Y(X) + H(Y)$,

(x) $H(X) - H_Y(X) = H(Y) - H_X(Y)$ (from (ix))

(xi) $H_X(Y) \leqslant H(Y)$, with equality if X and Y are independent.

(xii) $H(X,Y) \leqslant H(X) + H(Y)$ with equality as in (xi) (from (ix) and (xi)).

It can be shown[4], moreover, that properties (i), (iv), (viii) and (ix) uniquely characterize H (up to a positive constant). An alternative characterization can be given in terms of (ii), (viii) and (ix).[5]

By (iv), the maximum value of H for a given number m of alternatives is $\log m$. It is often convenient to describe one's uncertainty as the proportion of the maximal uncertainty obtainable with the same number of alternatives:

$$R = H/H_{max} = H/\log m .$$

The quantity R is called *relative entropy* (not to be confused with conditional entropy), and the complementary quantity $1 - R$ is called the *redundancy*.

If θ is a discrete parameter with possible values θ_i, we may define the *conditional sample information* (or *CSI*) associated with an outcome x_i of an experiment X by the reduction in entropy

$$H(\theta) - H_{x_i}(\theta) ,$$

which x_i effects, and then define the *expected sample information* (or *ESI*) yielded by X with the expected value of this quantity:

(1.14) $T(X;\theta) = H(\theta) - H_X(\theta)$,

a quantity (encountered above) which measures the average reduction in entropy which the outcomes of X effect. $T(X;\theta)$ is known to communications theorists as 'transmitted information'. This terminology is somewhat misleading, however, in its suggestion of a direction of flow. By property (x) above: $T(X;\theta) = T(\theta;X)$. Hence no direction of flow is implied: the information which the experiment transmits about the model is equal to the information which the model transmits about the experiment. $T(X;\theta)$ simply measures the association between two random variables. The symmetry of T is more apparent when it is written in the equivalent form:

(1.15) $T(X;\theta) = H(X) + H(\theta) - H(X,\theta)$.

Shannon's inequality, (xi) above, expresses the non-negativity of the *ESI*: $T(X;\theta) \geqslant 0$ with equality if X and θ are independent. We note the following straightforward extension of this inequality:

(1.16) $H(\theta) \geqslant H_X(\theta) \geqslant H_{XY}(\theta) \geqslant H_{XYZ}(\theta)$,   etc.

where, for example, $H_{XY}(\theta) = \Sigma_{ij}P(x_i, y_j)\Sigma_k P(\theta_k/x_i, y_j)$ is the uncertainty about θ when both X and Y are observed. Hence, the composite experiment (with outcome space $X \times Y$) is more informative, on the average, than either of its components. The information gained is a maximum when X and Y are independent, and a minimum when one is a sufficient statistic for the other. In fact, we have the following

THEOREM ON THE ESI. The *ESI* is (i) non-negative, (ii) additive in independent experiments, and (iii) preserved by a reduction $t(X)$ of the experimental random variable if $t(X)$ is a sufficient statistic.

Proof. (i) is Shannon's inequality; (ii) follows from the analogue of (xii) for conditional entropy, viz., $H_\theta(X, Y) = H_\theta(X) + H_\theta(Y)$, which is immediate from the extension of $P(X, Y) = P(X)P(Y)$ to $P(X, Y/\theta) = P(X/\theta)P(Y/\theta)$ when X and Y are independent. Then, using the symmetry of T, we have $T(X, Y; \theta) = T(\theta; X, Y) = H(X, Y) - H_\theta(X, Y) = H(X) + H(Y) - (H_\theta(X) + H_\theta(Y)) = T(\theta; X) + T(\theta; Y) = T(X; \theta) + T(Y; \theta)$. Now, for (iii), by the generalized Shannon inequality, $T(X, t(X); \theta) \geqslant T(t(X); \theta)$. But $T(X, t(X); \theta) = T(X; \theta)$, since observing X and $t(X)$ is the same as observing X. If t is sufficient, then the posterior distributions conditioned on x_i and on $t(x_i)$ are the same, so equality holds in $T(X; \theta) \geqslant T(t(X); \theta)$. If t is not sufficient, on the other hand, the posterior distributions based on x_i and on $t(x_i)$ are distinct for some i, and so the conditional entropies $H_{x_i}(\theta)$ and $H_{t(x_i)}(\theta)$ are distinct, and consequently, strict inequality holds in $T(X; \theta) \geqslant T(t(X); \theta)$.

4. NOISELESS INFORMATION

An experiment X (or channel) is *noiseless* for θ if $H_\theta(X) = 0$. I.e., knowledge of the true state or transmitted message removes all uncertainty regarding the outcome of X. Each outcome x of X may then be identified with the set of states θ such that x occurs when θ obtains, and so X is effectively a partition of θ, the set of states or possible messages. Consider now sequences of experiments or repetitions of an experiment where at each step there are n possible outcomes. Following Sneed (1967), I shall speak of *n-ary questioning procedures*. Given a noiseless channel, our problem may be to find the most efficient of all the *n*-ary questioning procedures (n is typically a function of the channel). It is not hard to see that the most efficient maximizes the *ESI* or transmitted information. (This maximum is often called the *channel capacity*.) For noiseless channels, $T(X; \theta) = T(\theta; X) = H(X) - H_\theta(X) = H(X)$. The best questioning procedure therefore maximizes $H(X)$, the outcome entropy, at each step. In particular, this procedure will identify the true state or message in a minimum number of steps, provided all partitions are feasible. In general, we must distinguish between the average number of steps it takes a procedure to identify the true state and the number of steps it *requires* to identify the true state. The latter is found by assuming the a priori least favorable distribution of states — the uniform distribution. For equiprobable messages, the best questioning procedure partitions the set of live possibilities into equinumerous subsets. I refer to this principle as the *uniform partition strategy*. For the problem of locating a square on a checkerboard discussed earlier, this strategy directs us to divide the number of remaining squares in half at each step. The following example further illustrates the efficiency of the uniform partition strategy.

EXAMPLE 5 (the odd ball). Given twelve steel balls, eleven of which are of the same weight, the problem is to locate the odd ball in three weighings with a pan balance, and to determine whether the odd ball is heavier or lighter than the eleven standard weights. (Thus, we seek a 3-ary questioning procedure that requires only three steps.) I number the balls $1, \ldots, 12$, and assign each of the 24 possible states $1H, 1L, 2H, 2L, \ldots, 12H, 12L$ ('H' for 'heavier', 'L' for 'lighter') equal probability. To insure noiselessness, I permit only weighings of equal numbers of balls. (Before reading on, the reader may wish to attempt a solution of this problem by trial and error.)

Solution. The uniform partition strategy determines the best first weighing as four against four (not, as many people initially guess, six against six). Say we weigh $1, 2, 3, 4$ against $5, 6, 7, 8$. Then all three possible outcomes of the weighing are equiprobable and the set of 24 possibilities is uniformly partitioned into three sets of 8 elements each. E.g., if the left pan is heavier, the unexcluded possibilities are $1H, 2H, 3H, 4H, 5L, 6L, 7L, 8L$. Given this outcome, let us find a best second weighing. Since there are 8 remaining possibilities, the best second weighing will partition this set into three subsets of $3, 3$ and 2 elements, the best feasible approximation to a uniform partition. Weighing $1, 2, 9$ against $3, 4, 5$ achieves this most nearly uniform partition, and is therefore a best second weighing. (N.B., 9 is known to be a standard weight.) Whatever outcome this best second weighing produces, the true state can be found on a third weighing. E.g., if the pans balance on the second weighing, leaving the possibilities $6L$, $7L$, $8L$, weigh ball 6 against ball 7. If they balance, you are left with $8L$, etc. The reader is invited to find a best second weighing in the case where the pans balance on the first weighing. Pursuit of the uniform partition strategy will yield the solution in three weighings whatever the outcome of each weighing. I.e., this questioning procedure *requires* only three questions.

An interesting new game, 'Mastermind', deserves mention in this connection. It is played with pegs of six colors. A code of four pegs is inserted by player I in holes not visible to player II, who must then decipher the code. At each move, player II places four pegs on the board, and player I tells him how many are of the right color, and how many of the right color are placed in the right column. Play is continued until player II cracks the code (say, (blue, red, green, red)). Although there are 6^4 possible codes and quite a few possible outcomes of each question, the uniform partition strategy still applies to determine the optimal move at any stage of play. Many simpler variants of the game, useful for pedagogical purposes, will readily suggest themselves.

Write 'H_n' for entropy when logarithms are to the base n. Let $N^*(X)$ be the average number of questions which it takes an optimal n-ary procedure X to identify the true state. Then, *provided all partitions are feasible*

$$(1.17) \qquad H_n \leqslant N^*(X) \leqslant H_n + 1 \; .$$

As Sneed (1967) observes, there is a one-one correspondence between separable n-ary codes and n-ary questioning procedures, and (1.17) is no more than a restatement of the so-called 'noiseless coding theorem'. In particular, writing $N(X)$ for the number of questions X requires, and noting that $H_n = \log_n m$ is the entropy of the source for m equiprobable messages, we have:

$$(1.18) \qquad \log_n m \leqslant N(X) \leqslant \log_n m + 1 \; .$$

so that $\log_n m$ gives a lower bound on the number of questions X requires to reach the true state and $\log_n m + 1$ an upper bound for an efficient procedure when all partitions are feasible.

In the checkerboard example, $n = 2$ (the questioning procedures are binary), there are $m = 64$ equiprobable states, $\log_2 64 = 6$, and six questions are required, so that the lower bound of (1.18) is attained. In the odd ball example, $m = 24$ and $\log_3 24 \leqslant N(X) \leqslant \log_3 24 + 1$ for the optimally efficient X, whence $N(X) = 3$, as we found above. If the number of balls were 36 (hence 72 possibilities), $\log_3 72 \leqslant N(X) \leqslant \log_3 72 + 1$ gives $N(X) = 4$, as the reader can verify. (You begin by weighing 12 against 12, and this reduces the number of possibilities to 24, a number which, we have found, is amenable to three weighings.) Note, though, when there are 13 balls and 26 possibilities, the most nearly uniform partition into subsets of 9, 9 and 8 possibilities is not feasible on the first weighing. (Weighing four against four gives the partition 8, 8, 10, while weighing five against five gives 10, 10, 6.) Hence, (1.18) does not apply, and, in fact, more than three weighings are required. Still, $\log_n m$ is seldom far off as an estimate of the number of questions required by an efficient n-ary procedure, and the same is true of the estimate H_n of the average number of questions which an efficient procedure uses to identify the true state. Thus, it hardly matters whether we use the estimate or the number it estimates as a measure of ignorance or of the disinterested information obtained by finding out the true state.

I have been comparing the efficiency of different n-ary questioning procedures given a fixed noiseless channel. The problem of comparing arbitrary noiseless experiments (i.e., arbitrary partitions of the state space) is much easier. Recall one partition is *as fine as* another if every member subset of the former is a subset of some member of the latter. This relation partially orders partitions, and we can say (as always) that one partition is finer than another if it is at least as fine and they are distinct. We carry over this standard terminology to noiseless experiments. The finest partition of a set is the collection of all single-membered subsets of that set, and the coarsest partition is the collection comprising the set itself as its only member. It should be obvious that a finer (noiseless) experiment has higher *ESI*. The following stronger result is due to Marshak and Radner (1972).

THEOREM ON FINENESS. Given any decision problem, one noiseless experiment has higher EVSI than another if it effects a finer partition of the state space.

The 'if' part is trivial; any utility attainable with the coarser partition is clearly attainable with the finer partition. The proof of the 'only if' part goes by constructing two payoff functions for a pair of partitions incomparable as to fineness such that the one has higher *EVSI* under the first payoff function and the other has higher *EVSI* under the second payoff function (for the details, see Marshak and Radner (1972), pp. 55–57).

The fineness condition is the weakest general sufficient condition that the *EVSI* be increasing in the *ESI*. That is, barring the important but exceptional case of two action two state problems, one can be sure that the

EVSI is increasing in the *ESI* of an experiment only if one experiment of a comparison has higher *ESI* by virtue of being finer.

EXAMPLE 6 (Marshak and Radner (1972), pp. 104–106). Let θ be the difference between the future and present value of a stock, and let θ be uniformly distributed in the closed interval $[-1, 1]$. Let e_n be the noiseless experiment which partitions the interval into n equal parts, $n = 1, 2, 3, \ldots$. If we assume that no more than one share can be purchased, then the amount exchanged is +1 or −1, depending on whether one share is bought or sold. The payoffs (which we identify with utilities) are given by $u(a,\theta) = a\theta$. Consider e_1 (the null experiment). Both actions have expected payoff zero, and so $EVSI(e_1) = 0$, the value of the prior Bayes act. Given the outcome of e_2, on the other hand, the speculator knows whether the price will rise (x) or fall (y), and he buys or sells accordingly. Writing x and y for the two outcomes of e_2, $a_x = 1$ and $a_y = -1$, and the posterior expected payoff of both these acts is $\frac{1}{2}$, the midway point between the best and worst possible payoff. Hence $CVSI(x) = CVSI(y) = \frac{1}{2} = EVSI(e_2)$. Denote the three possible outcomes of e_3, viz., θ in $[-1, \frac{1}{3})$, θ in $[-\frac{1}{3}, \frac{1}{3})$, θ in $(\frac{1}{3}, 1]$ by x, y, z, resp. Clearly, $a_x = -1, a_z = +1$, the posterior expected payoffs are both $\frac{2}{3}$, the midway point again, and hence $CVSI(x) = CVSI(z) = \frac{2}{3}$. On the other hand, $CVSI(y) = 0$, as with e_1, and so $EVSI(e_3) = (\frac{1}{3})(\frac{2}{3}) + (\frac{1}{3})0 + (\frac{1}{3})(\frac{2}{3}) = \frac{4}{9} < EVSI(e_2)$. It is clear, however, that using the appropriate extension of entropy to continuous variants, $ESI(e_3) > ESI(e_2)$. (Alternatively, we could replace the continuous intervals by suitably chosen discrete sets and obtain the same reversal.) The example turns on the fact that the disinterested information provided by outcome y of e_3 is without value for the decision at hand. It is worth more to distinguish between $\theta = -\frac{1}{4}$ and $\theta = +\frac{1}{4}$ (which can be done with e_2 but not with e_3) than to distinguish between $\frac{1}{5}$ and $\frac{2}{5}$ (which can be done with e_3 but not with e_2). The fact that the utility of given information varies from one decision problem to another is fraught with significance for those approaches to statistical hypothesis testing which purport to reduce inductive inference to decision making under uncertainty or 'inductive behavior' (cf. Chapter 10).

5. FISHER'S INFORMATION

In this section we study another measure of information (due to R. A. Fisher) which is convenient to apply when we wish to compare two experiments which yield estimates of the same parameter. Information about a parameter (or vector of parameters) θ can be measured, not only by entropy reduction, but by the precision of an estimate of J.

Consider first the case where θ is a single parameter. I follow here the elegant treatment of Edwards (1972) and write $L(\theta)$ for the likelihood function $p(x/\theta)$, and set $S(\theta) = \ln L(\theta)$. Edwards calls $S(\theta)$ the *support function*. The first derivative of $S(\theta)$ vanishes at the *ML* estimate $\hat{\theta}$; and so by expanding $S(\theta)$ about $\hat{\theta}$ in a Taylor series[6] we obtain:

(1.19) $\qquad S(\theta) = S(\hat{\theta}) + S''(\hat{\theta})(\theta - \hat{\theta})^2/2 + S'\,''(\hat{\theta})(\theta - \hat{\theta})^3/6 + \cdots$

Neglecting terms of higher than second order, (1.19) gives a parabolic approximation to the support curve near its maximum, with $-S''(\hat{\theta})$ the curvature. It seems appropriate to call $-S''(\hat{\theta})$ the observed information, writing $I(\hat{\theta}) = -S''(\hat{\theta})$, in as much as the difference in support between $\hat{\theta}$ and a nearby value is proportional to this quantity, and this difference intuitively reflects the informativeness of the experiment about θ.

As an example, consider the likelihood of a binomial parameter p, proportional to $p^a(1-p)^b$ when a successes and b failures are observed. Then $S(p) = a \ln p + b \ln(1-p)$, $S'(p) = a/p - b/(1-p)$, whence $\hat{p} = a/(a+b)$, the observed proportion of successes. Further, $-S''(\hat{p}) = a/p^2 + b/(1-p)^2$, whence, setting $a = n\hat{p}, b = n(1-\hat{p}), n = a + b$, we have $-S''(\hat{p}) = n/(\hat{p}(1-\hat{p}))$. The inverse square root of this quantity is called the standard error of $\hat{p}$. It is, of course, merely an approximation to the standard deviation of $\hat{p}$ obtained by substituting the estimate $\hat{p}$ for the unknown true value of p. Similarly, $-S''(\hat{p})$ is an approximation to the inverse variance of $\hat{p}$, a quantity called the *precision*. The same relation holds generally, and, since the true value of the parameter we are estimating is never known, we may, practically speaking, identify the observed information with the precision of an *ML* estimate. (See below for conditions under which this identification is exact.)

Let θ^* be the true value of θ. In what follows, expectations are taken with respect to the true data distribution, $p(X/\theta^*)$. Expecting both sides of the Taylor expansion (1.19) about θ^* gives:

$$E(S(\theta)) = E(S(\theta^*)) + E(S'(\theta^*))(\theta - \theta^*)$$
$$+ E(S''(\theta^*))(\theta - \theta)^2/2 + \cdots$$

Where differentiating under the integral sign is permissible, it is easily shown that $E(S'(\theta^*)) = 0$, and so we obtain the following analogue of (1.19)

(1.20) $\qquad E(S(\theta)) = E(S(\theta^*)) + E(S''(\theta^*))(\theta - \theta^*)^2/2 + \cdots.$

$E(S(\theta))$ is called the *mean support function*. It is the average of the different support curves associated with various experimental outcomes with respect to $p(X/\theta^*)$. For large samples, $\hat{\theta}$ will be close to θ^* and can be substituted for θ^* in (1.20).

Write L_n and S_n for the likelihood and support functions based on a random sample of n, so that $L_n(\theta) = p(x_1/\theta) \ldots p(x_n/\theta)$ and $S_n(\theta) = \ln L_n(\theta)$ is the sum of n independent and identically distributed random variables $\ln p(x_i/\theta)$, for each value of θ. By the strong law of large numbers, $\lim(S_n(\theta)/n) = E(S(\theta))$. Thus, to good approximation, the support at θ based on a random sample of n is n times a constant, viz., $E(S(\theta))$. Set $E(S(\theta)) = S^*(\theta)$. Using the Taylor expansion (1.20) with $\hat{\theta}$ in place of θ^*, and the fact that, barring rare exceptions, $S''(\hat{\theta})$ is negative, we obtain:

$$S^*(\theta) = a - c(\theta - \hat{\theta})^2 + h(\theta - \hat{\theta})^3 + \cdots .$$

Since, at large n, $S(\theta) = nS^*(\theta)$, we have

$$S(\theta) = na - nc(\theta - \hat{\theta})^2 + nh(\theta - \hat{\theta})^3 + \cdots .$$

The higher order terms in this expansion are negligible relative to the quadratic term. E.g., if $\theta - \hat{\theta}$ is of order $n^{-\frac{1}{2}}$, then the quadratic term is of order $n(n^{-\frac{1}{2}})^2 = 1$, while the cubic term is of order $n(n^{-\frac{1}{2}})^3 = n^{-\frac{1}{2}}$. In what follows, then, we may neglect these higher order terms to good approximation.

Suppose now that the prior density $p(\theta)$ is nowhere vanishing ('non-dogmatic'). Then the posterior density is proportional to $\exp(S(\theta) + \ln p(\theta))$. Since $S(\theta)$ is of order n, it eventually swamps $\ln p(\theta)$, and the posterior density is approximately proportional to

$$\exp(-(\theta - \hat{\theta})^2 / 2\sigma_n^2)$$

after absorbing $S(\hat{\theta})$ into the normalization constant and setting $\sigma_n^{-2} = -S''(\hat{\theta})$, and using, of course, the quadratic approximation to $S(\theta)$ given above. Hence, *the posterior density based on any non-dogmatic prior is asymptotically normal*. In particular, the likelihood function at large samples is of approximately normal shape. While the heuristic argument I have just sketched certainly makes these results plausible, no rigorous proof will be given here, and, in fact, the rigorous proof requires provisos concerning the existence and continuity of the relevant derivatives and their expectations. These conditions are ordinarily satisfied.[7]

The observed information is analogous to the conditional entropy, which measures uncertainty about the parameter of interest after observing the experimental outcome. For purposes of comparing experiments, however, the relevant quantity is the expected information. But how should we define it? Since the expected support function $E(S(\theta))$ has a maximum at θ^*, we measure the *expected information* by the negative curvature of the expected support function at this point, i.e., by $E(-S''(\theta)) = -E(S''(\theta))$. Just as the difference in support at $\hat{\theta}$ and a neighboring point is proportional to the observed information, the expected difference in support at neighboring points is proportional to the expected information. Note that the expected information, which I write $I^*(\theta) = -E(S''(\theta))$, is a function of the value θ of the unknown parameter; it tells us, therefore, how much information the experiment will yield on the average about the parameter of interest given that it assumes some particular value. Typically, neither of the two experiments will be more informative than the other for all values of θ, and so, if we want an overall comparison of two experiments, we must average $I^*(\theta)$ against our prior distribution of θ.

The definitions of observed and expected information extend to the multi-parameter case. Where θ and $\theta - \theta^*$ are column vectors, we have the Taylor expansion:

$$(1.21) \qquad S(\theta) = S(\theta^*) + (\theta - \theta^*)\left(\frac{\partial \theta}{\partial \theta_j}\right) + \frac{1}{2}(\theta - \theta^*)'\left(\frac{\partial^2 S}{\partial \theta_j \partial \theta_k}\right)(\theta - \theta^*)$$

$$+ \cdots$$

and again

$$(1.22) \qquad E\left(\frac{\partial S}{\partial \theta_j}\right) = 0$$

all partial derivatives being evaluated at θ^*.

The matrix whose general term is

$$(1.23) \qquad I_{jk} = -E\left(\frac{\partial^2 S}{\partial \theta_j \partial \theta_k}\right), \qquad \text{with partial derivatives evaluated at } \theta^*$$

is called *Fisher's information matrix*, or the *expected information matrix*. We have the alternative forms:

$$(1.24) \qquad -E\left(\frac{\partial^2 S}{\partial \theta_j \partial \theta_k}\right) = E\left(\frac{\partial S}{\partial \theta_j}\frac{\partial S}{\partial \theta_k}\right) = \int \frac{1}{L^*}\left(\frac{\partial L}{\partial \theta_j}\frac{\partial L}{\partial \theta_k}\right),$$

where $L^* = L(\theta^*)$. Provided the conditions for integrating under the integral sign are met, we have

$$\int \frac{\partial L}{\partial \theta_j} = 0 \quad \text{and} \quad \int \frac{\partial^2 L}{\partial \theta_j \partial \theta_k} = 0 ,$$

since $\int L = 1$, whence (1.24).

In particular, (1.24) holds for discrete sample spaces, integration giving way to summation. For a multinomial likelihood whose category probabilities $p_i = p_i(\theta_1, \theta_2, \ldots)$ are functions of one or more parameters θ_i, the expression on the right of (1.24) for the general term of the expected information reduces to

$$(1.25) \qquad I_{jk} = \Sigma_i \frac{1}{p_i} \left(\frac{\partial p_i}{\partial \theta_j} \frac{\partial p_i}{\partial \theta_k} \right).$$

Since the expected support function based on n independent observations is the sum of the respective support functions, the expected information is additive in independent experiments. Moreover, sufficient statistics are the only statistics which preserve expected information. In practice, of course, θ^* is unknown (considered now as a vector), and so one must evaluate partial derivatives at $\hat{\theta}$ in calculating the I_{jk}.

EXAMPLE 6 (linkage). In testing for genetic linkage, we often wish to compare the efficiency of different kinds of matings, e.g., the double backcross, $ABab \times abab$, with the single backcross, $ABab \times Abab$. The six possible genotypes of the offspring from these matings are assumed phenotypically distinguishable. The $ABab$ male parent produces gametes AB Ab, aB, and ab in the proportions $(1 - \theta)/2$, $\theta/2$, $\theta/2$, and $(1 - \theta)/2$, where the *recombinant types*, Ab and aB, result from crossing over, and θ, which is usually assumed to range from 0 to $\frac{1}{2}$ (though exceptional values $> \frac{1}{2}$ can occur), is called the *recombination fraction*. We have independent assortment (Mendel's case) when $\theta = \frac{1}{2}$, and highly proximate loci (genes which lie a small map distance apart on the same chromosome) for small values of θ. The female produces all ab gametes in the double backcross, and half Ab and half ab in the single backcross. The genotypic proportions expected from the two types of mating are given in Table VI. E.g., the genotype $Aabb$ can arise either when the male produces Ab or when the female produces Ab, and so the probability of this genotype is

$$\frac{\theta}{4} \frac{1}{2} + \frac{(1 - \theta)}{2} \frac{1}{2} = \frac{1}{4} .$$

TABLE VI

Genotype	Double backcross	Single backcross
AABb	0	$(1 - \theta)/4$
AAbb	0	$\theta/4$
AaBb	$(1 - \theta)/2$	$1/4$
Aabb	$\theta/2$	$1/4$
aaBb	$\theta/2$	$\theta/4$
aabb	$(1 - \theta)/2$	$(1 - \theta)/4$

I use (1.25) to compute the expected information from Table VI. For the double backcross, $(dp_i/d\theta)^2 = \frac{1}{4}$, whence $I^*(\theta) = 1/(\theta(1 - \theta))$. For the single backcross, (1.25) gives $I^*(\theta) = 1/(2\theta(1 - \theta))$, which is half the value of $I^*(\theta)$ for the double backcross. As Edwards (1972) notes (p. 149) in discussing this example, the results are not unexpected, since, in the double backcross, every cross-over in the male and none in the female can be detected — the situation is essentially binomial — while in the single backcross, a cross-over in the female parent is still undetectable, but in half the offspring (viz., *AaBb* and *Aabb*) the possibility that the *A* came from the mother makes it impossible to say whether or not a cross-over has occurred in the male. On the average, then, half of the offspring are uninformative, and the efficiency of the single backcross is just half that of the double backcross. This example is one of those rare cases where one experiment is uniformly more informative than the other; for cases of a similar sort — matings — where this does not happen, consult Mather (1951), Chapter VI.

If $\theta = f(\phi)$ is one-one, then $dS/d\phi = (dS/d\theta)(d\theta/d\phi)$ and $d^2 S/d\phi^2 = (d^2 S/d\theta^2)(d\theta/d\phi)^2 + (dS/d\theta)(d^2\theta/d\phi^2)$. Since $dS/d\theta = 0$ at the *ML* value of θ, the observed informations are related by

$$\frac{d^2 S}{d\phi^2} = \left(\frac{d\theta}{d\phi}\right)^2 \frac{d^2 S}{d\theta^2}$$

and the same is true of their expected values:

$$(1.26) \qquad I^*(\phi) = \left(\frac{d\theta}{d\phi}\right)^2 I^*(\theta).$$

There is an analogous relation between the expected information matrices in the multiparameter case.

We saw earlier that $E(dS/d\theta) = (d(\ln L)/d\theta)L \, dx = 0$. Let $f(\theta)$ be any parameter transformation, and let $t = t(X)$ be an unbiased estimator of $f(\theta)$, so that $E(t) = \int t\, L\, dx = f(\theta)$. Differentiating under the integral sign:

$$f'(\theta) = \int t\, \frac{dL}{d\theta}\, dx = \int t\, \frac{d(\ln L)}{d\theta} L\, dx = \int (t - f(\theta))\, \frac{d(\ln L)}{d\theta} L\, dx$$

and therefore, using the Cauchy-Schwarz inequality:

$$(1.27)\qquad (f'(\theta))^2 \leqslant \int (t - f(\theta))^2 L\, dx \int \left(\frac{d(\ln L)}{d\theta}\right)^2 L\, dx\,.$$

If $x = (x_1, \ldots, x_n)$ is a vector of observations, the integrals in (1.27) are, of course, multiple integrals. In the special case $f(\theta) = \theta, f'(\theta) = 1$, and (1.27) reduces to

$$(1.28)\qquad \text{var}(t) \geqslant 1/I(\theta)\,,$$

a result known as the *information inequality* (or the 'Cramer-Rao inequality'). The expected information thus imposes an upper bound on the precision of an estimator. An estimator which achieves this upper bound is called *efficient*. (Efficient estimators, as I am using the term, do not always exist.) The condition of equality in (1.27) is that of the Cauchy-Schwarz inequality, namely, that $t - f(\theta)$ be proportional to $d(\ln L)/d\theta$ for all observations x, or

$$(1.29)\qquad d(\ln L)/d\theta = A(t - f(\theta))\,,$$

where $A = A(\theta)$ may depend on θ. If (1.29) holds, the *ML* estimate satisfies $0 = S'(\hat{\theta}) = A(t - \hat{\theta})$, whence $t = \hat{\theta}$. Hence, *ML* estimators are efficient when efficient estimators exist. The latter are also easily shown to be sufficient.

As an example, let the likelihood L be normal with known variance, so that $\ln L = -(x - \theta)^2/2\sigma^2$. Then $d(\ln L)/d\theta = (x - \theta)^2/\sigma^2$ and $E(d(\ln L)/d\theta)^2 = \int ((x - \theta)^2/\sigma^4\, L\, dx = 1/\sigma^2$. The expected information for a random sample of n is therefore n/σ^2, and since $\text{var}(\bar{x}) = \sigma^2/n$, the sample mean $\bar{x}$ is an efficient estimator of the population mean θ. Notice, too, that $-d^2 S/d\theta = 1/\sigma^2$, so that the expected information and the observed information coincide, and either quantity gives the exact precision of the *ML* estimator. Huzurbazar (1949) has shown, more generally, that the exact precision of the *ML* estimator of a parameter θ is always equal to the observed (or expected) information when sufficient estimators exist (and, of course, when they do, they are *ML* estimators). The properties cited here are among those that lend *ML* estimation its theoretical importance.

In rounding out our discussion of Fisher's concept of information, I mention its relation to 'discrimination information' (Kullback, 1959). This measure applies when we are interested in comparing two hypotheses, H and K. The likelihood ratio, or its natural logarithm, $\ln(P(x/H):P(x/K))$, registers the import of the outcome x for this comparison. Good (1950) labelled this quantity *weight of evidence*, and its expected value, the *expected weight of evidence*. For discrete random variables we may write:

$$(1.30) \qquad I^*(H, K; X) = \Sigma_x P(x/H)\ln(P(x/H):P(x/K))$$

for expected weight of evidence, given that H obtains. Kullback (1959) calls (1.30) the *discrimination information* of X given H. I call the probability-weighted sum of $I^*(H\ K; X)$ and the symmetrically defined quantity $I^*(K, H; X)$, viz.,

$$(1.31) \qquad J^*(H, K; X) = P(H)I^*(H, K; X) + P(K)I^*(K, H; X)$$

the *discriminability* of H and K. I^* (hence, also, J^*) is easily shown to be non-negative, additive in independent experiments, and preserved by a reduction of the data iff it is a sufficient statistic (see Kullback, 1959, Chapter 2). For continuous variates, sums give way to integrals, as in the following example.

EXAMPLE 7. Given two normal populations $N(\mu_1, \sigma)$ and $N(\mu_2, \sigma)$ with common variance, the log likelihood ratio simplifies to

$$(1.32) \qquad P(x/\mu_1, \sigma) : P(x/\mu_2, \sigma) = \frac{\mu_1 - \mu_2}{\sigma} \frac{x - (\mu_1 - \mu_2)/2}{\sigma} ,$$

which is non-zero iff the means are distinct and the observation x does not fall midway between them. In that case, (1.32) increases with the separation of the means, measured in standard deviation units. By the symmetry of the problem, $I^*(1, 2) = I^*(2, 1)$, using obvious notation for the two hypotheses, and so $J^*(1, 2) = I^*(1, 2)$. Finally, letting $N(x/\mu_1, \sigma)$ denote the normal density:

$$I^*(1, 2) = \frac{\mu_1 - \mu_2}{\sigma} \int x N(x/\mu_1, \sigma)dx - \frac{\mu_1^2 - \mu_2^2}{2\sigma^2} \int N(x/\mu_1, \sigma)dx ,$$

$$= \mu_1(\mu_1 - \mu_2/\sigma^2) - (\mu_1^2 - \mu_2^2/2\sigma^2) ,$$

$$= (\mu_1 - \mu_2)^2/2\sigma^2 .$$

As expected, discriminability increases with the distance between the means measured in standard deviation units.

EXAMPLE 9. The *ESI* can be viewed as a special case of discrimination information. Given an experimental variate X and a parameter θ of some distribution, we compare the hypothesis H that X and θ are dependent, with joint density $p(x, \theta)$, with the hypothesis K that they are independent, with joint density $p(x)p(\theta)$. Then

$$
\begin{aligned}
I^*(H, K) &= \iint p(x, \theta) \ln\{p(x, \theta)/(p(x)p(\theta))\}\mathrm{d}x \, \mathrm{d}\theta \ , \\
&= \iint p(x, \theta) \ln p(\theta/x)\mathrm{d}x \, \mathrm{d}\theta + \iint p(x, \theta) \ln p(x)\mathrm{d}x \, \mathrm{d}\theta \\
&\quad - \iint p(x, \theta) \ln p(x)\mathrm{d}x \, \mathrm{d}\theta - \iint p(x, \theta) \ln p(\theta)\mathrm{d}x \, \mathrm{d}\theta \ , \\
&= \iint p(\theta/x) \ln p(\theta/x)p(x)\mathrm{d}x \, \mathrm{d}\theta - \iint \ln p(\theta)p(x, \theta)\mathrm{d}x \, \mathrm{d}\theta \ , \\
&= \int p(x)\{\int p(\theta/x) \ln p(\theta/x)\mathrm{d}\theta\}\mathrm{d}x - \int \ln p(\theta)\{\int p(x, \theta)\mathrm{d}x\}\mathrm{d}\theta \ , \\
&= -H_X(\theta) - (-H(\theta)) \ , \\
&= T(X; \theta) \ ,
\end{aligned}
$$

using the definition of entropy for continuous variates.

Since the expected information is approximately proportional to the expected difference in support at two neighboring parameter values, and the latter is given by $I^*(\theta_1, \theta_2)$, we should expect to have

$$(1.33) \qquad \lim I^*(\theta, \theta + \triangle\theta)/\triangle\theta^2 = I^*(\theta) \ ,$$

where the quantity on the left is the discrimination information (given θ), between θ and a nearby value $\theta + \triangle\theta$, and that on the right is the expected information at θ. A rigorous proof of this result is easily given using Taylor series (cf. Savage, 1954, p. 236); I omit the details.

Because I have compacted a good deal of material into this section, it would be well to summarize some of the main points.

(A) The expected information is an upper bound on the precision of an estimator; estimators which achieve this bound are called efficient. *ML* estimators are efficient when efficient estimators exist, and they are also sufficient.

(B) The (normalized) likelihood function is asymptotically normally distributed about the *ML* value $\hat{\theta}$ and with variance inverse to the expected information. In the multiparameter case, the normalized likelihood is asymptotically equal to a multivariate normal distribution with mean the vector of *ML* values and covariance matrix inverse to Fisher's information matrix.

(C) Given a one-one parameter transformation $\phi = f(\theta)$, the expected informations are related by $I^*(\phi) = I^*(\theta)(\mathrm{d}\theta/\mathrm{d}\phi)^2$.

(D) The 'disinterested' measures studied – the *ESI*, discrimination information, and Fisher's information – are all non-negative, additive in independent experiments, and preserved only by sufficient statistics.

6. CONCLUSION

Our discussion has by no means exhausted the measures of information that have been proposed. I have focused on what seem to me the most fundamental and most useful concepts, and on those which play a role in subsequent chapters. I have also wholly neglected the vast psychological literature dealing with applications of information theory to learning, perception and related problem areas. Much of this material is relevant to our concerns and highly suggestive, and so this is a serious omission. For a useful introduction to this literature, consult Atneave (1954), (1959), and Garner (1962). One would expect the 'disinterested' measures studied here to induce the same (or nearly the same) ranking of experiments, but I have not investigated the matter in detail (nor has anyone else, to my knowledge).

When we compare 'interested' with 'disinterested' measures, on the other hand, the matter is quite otherwise. The *EVSI*, we saw (Example 6), is not an increasing function of the *ESI* and the two can induce opposite rankings of the same pair of experiments. Consider another 'interested' measure (Blackwell and Girshick, 1954) which ranks one experiment higher than another if any loss function attainable with the latter is attainable with the former. As Lindley (1956) shows, one experiment ranked higher than another by this method must also have higher *ESI*, but the converse fails. Blackwell and Girshick show, for example, that in comparing the hypothesis that two traits F and G are unassociated with any alternative of dependence (where the proportions with which the two traits F, G occur in the general population are known), it is most informative to sample that one of the four traits F, G, non-F, non-G which is rarest in the considered population. This result can be verified directly for the *ESI*, and it follows from Lindley's more general result.

If utility one is assigned to the 'acceptance' of a true hypothesis and utility zero to the 'acceptance' of a false hypothesis, then expected loss reduces to the expected proportion of errors (i.e., of false accepted hypotheses). Lindley's result, seen in this light, is somewhat reassuring. On the other hand, as Marshak (1974) observes, if a_i is the action of affirming H_i, and we posit the 'disinterested' utilities $U(a_i, s_j) = \delta_{ij}$ (Kronecker's delta, which is 1 or 0 according as $i = j$ or $i \neq j$), then the *CVSI* of outcome x becomes

(1.34) $\max_i P(H_i/x) - \max_i P(H_i)$

as the reader can easily verify. However, not even this drastic constraint on the scientist's utilities will insure that the *EVSI* and *ESI* induce the same ranking of two or more experiments. One has only to note that the entropy of one distribution can exceed that of a second even though the maximal element of the first also exceeds the maximal element of the second. Thus, $H(1/8, 1/8, 3/4) > H(1/32, 8/32, 23/32)$ even though $3/4 > 23/32$. From this point of view, the information bearing on comparisons of sub-optimal hypotheses is without scientific interest.

In a similar vein, Hintikka and Pietarinen (1966) show that if a scientist is regarded as a decision maker interested only in one utility, viz., information, as measured by $\text{Cont}(H) = 1 - P(H)$, then the expected utility of affirming H is given by

(1.35) $P(H/x)\text{Cont}(H) - P(-H/x)\text{Cont}(-H)$,

which simplifies to

(1.36) $P(H/x) - P(H)$.

The increment $P(H/x) - P(H)$ can be taken to measure the *degree of confirmation* which x affords H. From this point of view, then, the scientist is viewed as interested in maximizing degrees of confirmation, and this would lead him to rank or compare experiments by the measure

(1.37) $\Sigma_x P(x)\max_H (P(H/x) - P(H))$,

the summation being over all experimental outcomes x.

NOTES

[1] Here I am assuming that the utilities of act-state pairs and of sampling costs are separable: $U(e, x, a, s) = U(a, s) + U(e, x)$, where $U(e, x, a, s)$ is the utility of performing e, observing x, and doing a when s is the state. This assumption is violated, for example, if the utility of doing a when s obtains depends on which outcome x is observed.
[2] We allow X to assume countably many values x_i, $i = 1, 2, 3, \ldots$.
[3] Sums give way to integrals in the most straightforward extension of the entropy concept to continuous variates. But entropy, so defined, is not invariant under parameter transformations. Jaynes (1968) shows how to extend entropy so as to insure invariance.
[4] Khinchin (1957), pp. 9ff.
[5] Shannon and Weaver (1949), pp. 49ff.

[6] I am assuming here (and in what follows) the existence of derivatives and partial derivatives of all necessary orders. See Edwards (1972), Chapter 8 for a treatment of anomalous cases.

[7] A rigorous proof mimics that of the analogous result that $\hat{\theta}$ (the *ML* estimator) is asymptotically normally distributed about θ^*, the true value; see Cramer (1946), Section 33.2 for the proof and a statement of the conditions under which the theorem holds. Lindley (1965), p. 140, shows that the theorem fails for $p(x_i/\theta) = \theta^{-1}$ for $0 \leqslant x_i \leqslant \theta$, and $p(x_i/\theta) = 0$ otherwise. The posterior density for a sample of n is then approximately proportional to θ^{-n}, and this clearly does not tend to normality.

BIBLIOGRAPHY

Adams, E. W.: 1966, 'On the Nature and Purpose of Measurement', *Synthese* 16, 125–168.

Atneave, F. J., 1954, 'Some Informational Aspects of Visual Perception', *Psych. Rev.* 61, 183–193.

Atneave, F. J.: 1959, *Applications of Information Theory to Psychology*, Holt, Rinehart, and Winston, New York.

Blackwell, D. and Girshick, M. A.: 1954, *Theory of Games and Statistical Decisions*, Wiley, New York.

Braga-Illa, A.: 1964, 'A Simple Approach to the Bayes Choice Criterion: The Method of Extreme Probabilities', *J. Amer. Stat. Ass.* 59, 1227–1230.

Cramer, H.: 1946, *Methods of Mathematical Statistics*, Princeton University Press, Princeton, New Jersey.

Edwards, A. W. F.: *Likelihood*, Cambridge University Press, Cambridge.

Fisher, R. A.: 1922, 'On the Foundations of Theoretical Statistics', *Phil. Trans. Roy. Soc. London* A222, 309–368.

Garner, W. R.: 1962, *Uncertainty and Structure as Psychological Concepts*, Wiley, N.Y.

Good, I. J.: 1950, *Probability and the Weighing of Evidence*, Griffin, London.

Good, I. J.: 1956, 'Some Terminology and Notation in Information Theory', *Proc. I.E.E.*, Part C, pp. 103, 200–204.

Goosens, W. K.: 1970, *The Logic of Experimentation*, Doctoral Dissertation, Stanford University.

Hintikka, K. J. J., and Pietarinen, J.: 1966, 'Semantic Information and Inductive Logic', in K. J. J. Hintikka and P. Suppes (eds.), *Aspects of Inductive Logic*, North-Holland Publ. Co., Amsterdam.

Huzurbazar, V. S.: 1949, 'On a Property of Distributions Admitting Sufficient Statistics', *Biometrika* 36, 71–74.

Khinchin, A. I.: 1957, *Mathematical Foundations of Information Theory*, transl. from the Russian by R. A. Silverman and M. D. Friedman, Dover, N.Y.

Kullback, S.: 1959, *Information Theory and Statistics*, Wiley, N.Y. (reprinted: Dover, N.Y., 1968).

Lindley, D. V.: 1956, 'On a Measure of Information Provided by an Experiment', *Ann. Math. Stat.* 29, 986–1005.

Lindley, D. V.: 1965, *Introduction to Probability and Statistics*, Vol. 2, Cambridge University Press, Cambridge.

Marschak, J. and Radner, M.: 1972, *Economic Theory of Teams*, Yale University Press, New Haven.

Marschak, J.: 1974, 'Prior and Posterior Probability and Semantic Information', in G. Menges (ed.), *Information, Inference and Decision*, D. Reidel Publ. Co., Dordrecht.

Mather, K.: 1951, *The Measurement of Linkage in Heredity*, Methuen, London.
Mosteller, F.: 1965, *Fifty Challenging Problems in Probability with Solutions*, Addison-Wesley, Reading, Mass.
Raiffa, H. and Schlaifer, R.: 1961, *Applied Statistical Decision Theory*, Harvard Business School, Boston.
Raiffa, H.: 1968, *Decision Analysis*, Addison-Wesley, Reading, Mass.
Reza, F. M.: 1961, *An Introduction to Information Theory*, McGraw-Hill, New York.
Rosenkrantz, R. D.: 1970, 'Experimentation as Communication with Nature', in K. J. J. Hintikka and P. Suppes (eds.), *Information and Inference*, D. Reidel Publ. Co., Dordrecht.
Savage, L. J.: 1954, *Foundations of Statistics*, Wiley, New York.
Shannon, C. E. and Weaver, W.: 1949, *The Mathematical Theory of Communication*, University of Illinois Press, Urbana, Ill.
Smith, C. A. B.: 1967, *Biomathematics*, Vol. 2, Hafner Publ. Co., New York.
Sneed, J.: 1966, 'Entropy, Information and Decision', *Synthese* 17, 392–407.

CHAPTER 2

THE PARADOXES OF CONFIRMATION

1. GENERAL LAWS

There are those who think philosophers have already spilled more ink on the paradoxes of confirmation than they are worth, others who think them among the deepest conceptual knots in the foundations of knowledge. Like the problem of free will, the Goodman paradox owes much of its fascination to the way in which it combines urgent and topical philosophical concerns, above all, interest in the inductive roots of language and the linguistic roots of our theoretical construction of the world. My own view is that the paradoxes have at least one lesson, fraught with significance, to convey: *that general laws are not necessarily confirmed by their positive cases.* I. J. Good[1] was, I believe, the first to both point this out and make a convincing case. One of his examples is rather artificial. He invites us to imagine that the live possibilities have been narrowed to just two: either the world contains a single white raven and a vast number of black ravens, or else it contains no white ravens and a modest number of black ravens. (In either case, of course, it may contain other things as well.) Since a random sample of the general population is more likely to contain a black raven in the first case than in the second, the first possibility is confirmed (i.e., made more probable). Hence, by confirming the first possibility (many black ravens and a single white raven), observation of a black raven disconfirms the hypothesis that *all* ravens are black.

Good's other example is less precise but more suggestive. Crows and ravens being related species, observation of white crows (mutants, perhaps) would tend rather to lower than raise the probability that all ravens (including mutants) are black. This example provides more insight. When we fill out the description of a non-black non-raven to include its being a white crow, *we bring relevant background knowledge into the foreground*. The same dramatic affect on our probabilities is illustrated in a somewhat sharper form when we fill out our description of a non-reactive specimen of non U^{238} (the heavy isotope of uranium) to include its being an inert specimen of U^{235} (the lighter isotope). Atomic theory instructs us that the chemical properties of an element are independent of isotopy, and so we should hardly expect observation of inert specimens of the lighter isotope of uranium to increase our confidence that samples of the heavier isotope are reactive. There is an even better example to illustrate the point, one which has the virtue of bringing knowable probabilities into play.

Consider the classical problem of *matches*.[2] A case would be assorting N hats at random among their N owners; the problem is to compute the probability of a match (a man receiving his own hat). Let H be the hypothesis that no man receives his own hat (no matches). Of the first two men queried, we learn that neither received his own hat (in conformity with H). This outcome, call it X, will confirm H. But let us see what happens when we pick out various subevents of X.

Let E be the event that neither of the first two men receives his own hat *or the other's*, F the event that neither receives his own hat and *one* receives the other's and G the event that neither receives his own and *each* receives the other's. All three of these mutually exclusive and jointly exhaustive sub-events of X conform with H, but clearly, they bear quite differently on the probability of H. Suppose there are $N = 7$ men in all. When $N = 7$, $P(H) = 0.3679$; when $N = 5$, $P(H) = 0.3667$, which is smaller. But $P(H/G)$ is just the probability that no man receives his own hat when $N = 5$, and so G *disconfirms* H. In fact, all the relevant calculations for this problem can be neatly summarized thus:

$$P(H/E) = 94/120, \quad P(H/F) = 61/120, \quad P(H/G) = 44/120 ,$$

$$P(E/X) = 20/31, \quad P(F/X) = 10/31, \quad P(G/X) = 1/31 ,$$

from which

$$P(H/X) = P(H/E)P(E/X) + P(H/F)P(F/X) + P(H/G)P(G/X)$$

$$= 0.6812 .$$

Consequently, G disconfirms H, F weakly confirms H, and E strongly confirms H. Since the posterior probability of H given X is a weighted average of its posterior probabilities given E, F, and G, the confirmation afforded H by X will likewise be a weighted average of the degrees of confirmation afforded H by E, F, and G. Hence, X moderately confirms H (more weakly than E and more strongly than F). It makes all the difference, therefore, which subevent of X we single out. For the sake of completeness, I sketch the elementary combinatorics which lead to the given probabilities.

How many assortments realize X (neither 1 nor 2 receives his own hat)? If individual 1 receives the hat of 3, 4, 5, 6 or 7, then the hat assigned 2 can be chosen in 5 ways, since 2 cannot receive his own — whence 25 possibilities — while if 1 receives 2's hat, the hat assigned 2 can be chosen in 6 ways. The hats of 1 and 2 having been assigned, the remaining 5 hats can be assorted among the other 5 men in 5! = 120 ways. Hence 31 x 5! assortments realize X. Next consider E (neither 1 nor 2 receives his own hat or the other's hat). In this case, the hats assigned 1, 2 are chosen from the hats of 3, 4, 5, 6, 7 in $\binom{5}{2}$ = 10 ways, and the chosen two can be assigned to 1, 2 in either of two orders. Again the remaining 5 hats can be assorted in 5! ways. Hence 20 x 5! assortments realize E and $P(E/X) = 20/31$. Clearly, 5! assortments realize G, whence $P(G/X) = 1/31$. By subtraction, $P(F/X) = 10/31$.

Consider now $P(H/E)$. E itself splits into subevents: e.g., 1, 2 can receive the hats of 3, 4 or of 3, 5, etc. But it is only necessary to consider one such subevent, say 3, 4, since the probability of H is the same conditional on any of them. Hats 3, 4 having been assigned 1, 2 we count the number of ways H can be violated. First, only 5, 6 or 7 can receive his own hat, since the hats of 3, 4 are already assigned to 1, 2. The event that each of 5, 6, 7 can receive his own hat can happen 2 ways, that two of the three receive their own hat can happen in 4 ways (making 12 new possibilities), while the event that exactly one of the three receives his own hat can happen 4 ways (making another 12 possibilities). Hence, H can be violated in 26 ways when 1, 2 receive the hats of 3, 4. Since the 5 hats assigned to 3, 4, 5, 6, 7 can be assorted in 5! ways altogether, $P(H/E) = 1 - 26/120 = 94/120$. One arrives at $P(H/F)$ and $P(H/G)$ in the same way.

2. RESOLUTION OF THE PARADOXES

The lynch pin of the Goodman paradox is the inference from 'a green emerald examined before time t is a grue emerald' to 'examination of an emerald before time t which proves to be green confirms the hypothesis that all emeralds are grue'. But when we fill out our description of a grue emerald to include its being green and examined prior to time t, we single out a subevent, and no inference to the confirmation of the grue hypothesis can be drawn. No more than we can infer confirmation of the reactivity of the heavy isotope of uranium by inert specimens of the light isotope from the fact that the latter are non-samples of the heavy isotope.

Nor, for that matter, can we even infer confirmation of the grue hypothesis by observation of grue emeralds, for, in general, as Good's first example illustrates, we cannot conclude confirmation of 'All A are B' from observation of AB's. The possible worlds (i.e., the possible states of the actual world with respect to a specific population and set of properties) which are assigned high prior probability in the light of background knowledge and contain many AB's may all contain an A which is non-B. Just as in Good's example, finding an AB would, by raising the probabilities of these possible worlds, lower the probability that all A are without exception B.

The same is true, *a fortiori*, of non-A B's. In fact, it is quite easy to think up cases where observation of a non-A B would disconfirm 'All A are B'. This would be true, for example, if the numbers of A's and B's were known and finite. (For 'All A are B' to have non-zero prior probability would then require that the known number of B's exceed the known number of A's.) Each non-A B found would then reduce the probability that all A are B, the probability vanishing entirely when the observed number of non-A B's surpassed the known excess of B's over A's.

When background knowledge is admitted, e.g., in the form of a probability distribution over possible states of the considered population, very little can be inferred in general about the confirmation of a general law by its 'positive cases' (in any straightforward sense of this term). Given a probabilistic analysis of confirmation, then, the paradoxes are stopped dead in their tracks. We cannot infer the confirmation of the grue hypothesis by grue emeralds (much less by green emeralds), nor that of the raven hypothesis by white shoes or red herrings.

There is, however, one weak sort of exception to the above. Following Good, let us imagine that a stooge records the outcome of the experiment and reports only whether or not the outcome was consistent with the general law tested. Good calls such observations *stoogian*. In the hat example, the outcome that neither of the first two men queried received his own hat would be 'stoogian' in the indicated sense. That outcome, we saw, did (weakly) confirm the hypothesis that no man received his own hat. This example suggests what is true in general: *that a conforming stoogian observation is confirming* (or, at any rate, not disconfirming). For let X be a conforming stoogian outcome for H. Then $P(X/H) = 1$, since, if H holds, the outcome must be consistent with H. Hence, unless $P(X) = 1$ (which would mean

$P(X/\text{not} -H) = 1$), $P(X/H):P(X) > 1$, and so, by Bayes' rule, $P(H/X)$ exceeds $P(H)$.

3. LAWLIKENESS

In Professor Goodman's view, as I understand it, the Hempelian (satisfaction) criterion of confirmation is sound, provided that it is restricted to lawlike hypotheses.[3] While, informally, one might characterize as 'law-like' those hypotheses which are confirmed by their Hempelian instances, an independent criterion must be provided. Roughly, Goodman proposes to measure lawlikeness by the entrenchment of the predicates involved. Entrenchment reflects the frequency with which the extension of a predicate has been projected or has figured in 'successful' inductions. (It is in this form, Goodman holds, that our past inductive experience bears on our present practice.) I must assume in what follows that at least the main outlines of Goodman's entrenchment theory are familiar to the reader.[4]

My own view is that entrenchment *per se* has but little to do with confirmability or lawlikeness. What matters is, not that the predicates of an hypothesis have figured in successful laws or theories, but that there be theoretically grounded relations between the predicates which the hypothesis links. And that is only to say the residual background theory provides some prior presumption in favor of the hypothesis.

Whatever else it is intended to do, the entrenchment theory is aimed at driving a wedge between the grue hypothesis and its presumably non-pathological counterpart, 'All emeralds are green'. While 'green' and other ordinary color terms have figured in many a good empirical law, 'grue' has not. Yet, it is not hard to cook up cases, given the right sort of background knowledge, where a gruelike hypothesis is both more confirmable and enjoys higher prior probability.

Consider a twofold classification of chips in an urn: they are circles or squares, black or white. 'Blite' applies to black squares or white circles. Assume all 2^{2^N} states of the population of N chips equiprobable. Then I claim the hypothesis K that all chips are blite is more confirmable than the hypothesis H that all circular chips are black. Here 'more confirmable' means 'better confirmed by a conforming outcome'. Suppose that a random sample of one yields a black square, an outcome which conforms to both hypotheses. It can be shown this outcome is more probable on K than on H, and hence,

1   2   3   4   5   6   7   8   9 10 11 12 13 14 15 16

○   ■   ○   □   ○   ■   ■   □   □   ●   ●   ○   ■   ●   □   ●

○   ○   ■   ○   □   ■   □   ■   □   ●   ○   ●   ●   ■   ●   □

Fig. 1.

better confirms K. The case $N = 2$ is shown in Figure 1. Let X be the outcome in question (drawing a black square). We have $P(X/K) = (1 + 1/2 + 1/2 + 0)/4 = 1/2$, while $P(X/H) = (1 + 1/2 + 1/2 + 1/2 + 1/2 + 0 + 0 + 0 + 0)/9 = 1/3$, there being 9 states in which H holds (6–10, 13–16) and just 4 in which K holds (1–3, 6 in Figure 1).

By assigning states 1–3, 6 high prior probabilities, one could also make K initially more probable than H without making it less confirmable. In any case, confirmability and prior probability must be distinguished. Finally, let X be the conformable stoogian observation for H and Y that for K. Clearly, $P(X) > P(Y)$, so that the *a priori* chance of an outcome conforming to K is smaller, and so, by Bayes' rule, K is more strongly confirmed by a conforming stoogian observation than H.

Industrial melanism[5] illustrates the situation depicted here. In industrially polluted regions of Great Britain, little lichen grows on the trees, and against the background of the darker lichen-free trunks, dark-colored moths are less visible to predators and enjoy a selective advantage. The reverse is true of the heavily lichen-covered trees characteristic of rural, unpolluted areas. Let 'blite' apply to black (or dark) moths inhabiting polluted regions and to white (or light) moths inhabiting unpolluted regions. Then, given evolutionary theory, 'Most British moths are blite' is more confirmable and more probable than 'Most British moths are white', where 'most' means 'all save rare mutants'.

There is, then, nothing so very paradoxical in the suggestion that a stoogian observation consistent with a gruelike hypothesis should confirm it. Goodman's paradox has its sting rather in the claim that observation of green emeralds before time t confirms the grue hypothesis, and all the more so when this is coupled with the tacit suggestion that, because grue emeralds found after t are blue, finding green emeralds before t increases the probability of finding blue emeralds after t. Even if the first inference went through (which it does not), the second would require a sort of transitivity principle or consequence principle (whatever confirms H confirms any consequence of H). But it is child's play to find examples where X increases the probability of *H* & *K* but decreases the probability of *K*, or where *X* confirms *H* while disconfirming a consequence of *H*.[6]

Must we then conclude, in the light of the foregoing considerations, that no trait of hypotheses themselves makes for greater confirmability? Actually, that is not my view. I think, rather, that confirmability, while assuredly dependent on background knowledge, is also a function of the simplicity or specificity of the hypothesis (relative to the contemplated experiment). That is, the fewer possible experimental outcomes an hypothesis or theory fits, the more plausibility it gains when the outcome actually observed is found to fit. One can see, in a rough and ready way, why this is so by considering stoogian observations. The fewer possible outcomes an hypothesis fits, the smaller the probability of a conforming stoogian outcome *X*, and so the greater the ratio $P(X/H) : P(X)$ by which *X* multiplies the prior probability. The thesis that simpler theories are more confirmable is really the backbone of this book, and will occupy us from Chapter 4 on.

4. CONCLUSION

Professor Hempel has insisted from the start that his definition of confirmation in terms of satisfaction applies only to 'theoretically barren' contexts where background knowledge is lacking. If that proviso is respected, we can no longer claim, for example, that finding white crows actually disconfirms the raven hypothesis, for there we were appealing to the knowledge that crows and ravens are biologically related species. On the other hand, if we suppressed our belief that emeralds in all parts of space-time are the same color, it would be hard to see what is paradoxical about the alleged confirmation of the grue hypothesis by green emeralds. Be that as it may, let us consider Hempel's restriction.

It can be shown[7] that if all compositions of a doubly dichotomized population are equiprobable, then all three of the conforming types, *AB*'s, non-*A B*'s, and non-*A* non-*B*'s, afford 'All *A* are *B*' equal confirmation (in a probability sense). For then, as is evident from symmetry, all three types are equiprobable conditional on the hypothesis in question. If we equate then a 'barren' context with a uniform distribution over worlds (state descriptions), we can affirm a strong form of Hempel's claim that, appearances notwithstanding, non-*A*'s are genuinely confirmatory instances of 'All *A* are *B*'. Without some such condition, however, we cannot even affirm that *AB*'s are confirmatory.

On the other hand, if by a 'barren' context is meant one in which no pertinent background knowledge at all is available, I would be inclined to doubt whether such contexts exist. The very structure of human perceptual systems, with their built-in tendencies to group objects, select given features, or reidentify features, may already be tantamount to background knowledge. Certainly the use of terms which denote natural kinds already presupposes a good deal. Even if a clear meaning could be given to the conception of inductive inference in a theoretical vacuum, I do not see how anything could be concluded about confirmation. In short, the real issue between Hempel and the Bayesians, as I see it, is whether anything at all can be inferred about confirmation on the basis of the semantical relation of satisfaction alone, beyond what I have already noted: that conformable stoogian observations are confirming. The significance of these reflections for the traditional problem of 'justifying induction' will be taken up next.

NOTES

[1] See Good (1967).

[2] For mathematical details, consult Feller (1957), pp. 97–99.

In his *System of Logic*, Bk. III, Chapter III, Section 3, J. S. Mill asks: "Why is a single instance, in some cases, sufficient for a complete induction, while in others, myriads of concurring instances, without a single exception known or presumed, go such a very little way towards establishing a general proposition?" And he adds, "whoever can answer this question . . . has solved the problem of induction".

[3] The original statement of the theory in Goodman (1965), Chapter 4 will suffice for our purposes.

[4] 'Further Selection Experiments on Industrial Melanism in the *Lepidoptera*', H. B. D. Kettlewell, *Heredity* **10** (1956), 287–301.

[5] See Rosenkrantz (1973), pp. 157–158 for examples. It is also shown there that if we measure degree of confirmation by the plausibility increment, $P(H/E) - P(H)$, that the degree of confirmation a consequence E of H affords the conjunction $H\&K$ is given by:

$$dc(H\&K, E) = P(K/H)[P(H/E) - P(H)] = P(K/H)dc(H, E),$$

i.e., the dc E affords the conjunction $H\&K$ is the fraction $P(K/H)$ of the dc E affords H alone. Now the grue hypothesis is the conjunction of 'All emeralds examined prior to t are green' (H) with 'All emeralds not examined prior to t are blue' (K), and since $P(K/H)$ is presumably minute, the dc a conforming stoogian observation E for H affords $H\&K$, the grue hypothesis, is a miniscule fraction of the dc E affords H alone. But, equally, it might be objected, 'All emeralds are green' is the conjunction of 'All emeralds examined prior to t are grue' (H') with 'All emeralds not examined prior to t are bleen' (K'). Notice, however, that $P(K'/H')$ is the probability that emeralds unexamined prior to time t are the *same color* (i.e., reflect light waves of the same length) as emeralds examined prior to t. Hence, $P(K'/H')$ is much larger than $P(K/H)$, at least, given the background knowledge we do in fact possess. Here, then, is an asymmetry that does allow us to conclude that a conforming stoogian observation affords the 'green' hypothesis higher dc than a conforming stoogian observation accords the 'gruesome' hypothesis.

[7] Rosenkrantz (1971), p. 194.

BIBLIOGRAPHY

Feller, W.: 1957, *Introduction to Probability Theory and its Applications*, Wiley, New York.

Good, I. J.: 1967, 'The White Shoe Is a Red Herring', *Brit. J. Phil. Sci.* 17, 322.

Goodman, N.: 1965, *Fact, Fiction and Forecast*, Bobbs-Merrill, Indianapolis, Ind.

Hempel, C. G.: 1965, *Aspects of Scientific Explanation*, Chapter 1, The Free Press, New York.

Rosenkrantz, R. D.: 1971, 'Inductivism and Probabilism', *Synthese* 23, 167–205.

Rosenkrantz, R. D.: 1973, 'Probabilistic Confirmation Theory and the Goodman Paradox', *Am. Phil. Quart.* 10, 157–162.

CHAPTER 3

INDUCTIVISM AND PROBABILISM

1. THE INDUCTION RULE

It is considered a truism that not all inferences are deductive. Arguments which compel assent, granted their premises, are said to be limiting cases of the more general class of those whose premises support their conclusions inconclusively. Such arguments are variously labelled 'inductive', 'informative', or 'probable'. The conclusions of deductive arguments from true premises are invariably true, while conclusions of probable arguments from true premises are said to be 'for the most part true'.[1] One speaks also of 'arguments by analogy', 'statistical syllogisms', and the like – terms which further argue that induction and deduction are but two sides of the same inferential coin.

Whatever the rubric preferred, non-deductive inferences have notoriously eluded precise characterisation. In particular, no clear-cut analogue of the semantical concept of validity has emerged for inductive arguments. This, despite heroic attempts of Carnap, Hempel and others to extend the deductive concept of implication to an inductive concept of partial implication or confirmation.[2]

Quite distinguishable from the work of the Carnapians, an older tradition persists which grounds all non-demonstrative inference on a single principle, the induction rule. Classic formulations of it assert that events conjoined in the past are likely to be conjoined in the future, that like causes have like effects, or that generalizations are the more probable the greater the number of their attested cases. Under the first of these formulations, the principle constitutes an organon of discovery, but it has also been thought to provide a rule of estimation (Reichenbach's 'straight rule'), a measure of confirmation (as in the third formulation), and a method of computing predictive probabilities. Only the foundation upon which it rests has been regarded as obscure and standing in need of elucidation. This is the celebrated problem of justifying induction. Otherwise put, the problem is to explain why 'inductive methods' work. But it is not clear just what has worked, historically or comparatively speaking.

The corpus of inductive methods can be pinned down somewhat in terms of the induction rule. But that leaves one no place to go so far as justification is concerned. For if the rule is presupposed in the extrapolation of its own successful performance, then we are back where we began – in the Humean bind. It was perhaps inevitable that someone should eventually make a merit of this 'scandal', and declare the very impossibility of justifying or repudiating the rule without pre-supposing it as constituting, if not justification, then 'vindication'.[3] Related attempts to unravel Hume's knot have argued justification is uncalled for, that "it is an analytic proposition . . . that . . . the evidence for a generalization is strong in proportion as the number of favorable instances . . . is great".[4] To ask whether it is reasonable to place reliance on the induction rule, the attempted dissolution continues, "is like asking whether it is reasonable to proportion the degree of one's convictions to the strength of the evidence".[4]

But while it is senseless to demand rational justification of the rational standards of justification, it does not follow that the induction rule is one among, much less the fundamental one among, the canons of inductive rationality. This is the very point that should be at issue. For there is ample reason to cast a wary eye at the grandiose claims made on behalf of the rule.

The paradoxes of confirmation have exposed the problematic character of instancehood. Given a generalization 'All A are B', it is clear that $A\bar{B}$'s are negative instances. But not much else is clear. Most inductivists countenance AB's as positive instances, but if one also accepts the equivalence condition, then $\bar{A}\bar{B}$'s and $\bar{A}B$'s must also be admitted as positive instances. In this way we are led to the conclusion that everything is relevant, and this, I contend, is absurd on the face of it – especially when combined with a requirement of total evidence.[5] Indeed, under this naive conception of instancehood, far from being analytic, the induction rule is just false (see Chapter 2). Goodman's closely related paradox also casts doubt on the thesis that everything consistent with an hypothesis confirms it, but, in addition, it draws our attention to the varying rates at which confidence in an hypothesis grows in diverse theoretical contexts. Nor is it obvious how to account for this salient feature of inductive reasoning in terms of the rule. For the latter is quite literally 'enumerative' in its suggestion that we can assess evidence merely by counting instances.

In addition to these familiar difficulties, there are others, equally serious, that have received less attention. For one thing, it has never been made clear how the induction rule is to account for the support of causal and statistical hypotheses. Having observed 21 cures in 25 innoculations of a new drug, should I place more or less confidence in its efficacy than had I observed 82 cures in 100 trials? The induction rule by itself provides no clue how to weight the two factors involved in such appraisals: the size and the definitiveness of the sample. In point of fact, the dogma that the induction rule lies at the heart of all our experimental inferences masks a host of similar difficulties and draws attention away from the essential ambiguity of the rule itself.

2. THE IMPORT OF DATA

Classical formulations omit reference to background knowledge and so pose a spurious optimization problem: there is no optimal inductive method that depends only on frequency counts. For frequency counts have no import whatever in abstraction from a probability model. Broadly speaking, inductive reasoning allows us to base conclusions about the future behavior of a process on its past behavior. But the import of data about the past for the future is a function of the laws in accordance with which the process develops.

Polya urn models nicely illustrate this point. Imagine an urn containing b black and r red balls. A ball is drawn at random and replaced. But in addition, c balls of the color drawn and d balls of the opposite color are placed in the urn. A new drawing is made from the urn, now containing $b + r + c + d$ balls, and the procedure repeated.[6] In Polya's original scheme, $d = 0$ and $c > 0$, so that the drawing of either color increases the probability that the same color will be drawn next. This gives a rough model of contagion, where the quotient $c/(b + r)$ measures the rate of contagion. Other bounds on the parameters lead to different models. E.g., $c = -1, d = 0$, defines the model of random sampling without replacement, terminating after $b + r$ steps.

Now it should be clear that a given relative frequency of black balls on past drawings from a Polya urn can connote either an increase or a decrease in the probability of drawing a black ball on the ensuing trial, depending on the specifications of the parameters b, r, c, d. Nevertheless, by the lights of the induction rule, an abstract pattern of trial outcomes, a binary sequence, would have the same import for the future development of a process evincing fatigue as it would for a contagious process. For the trial outcomes are all that the induction rule takes explicitly into account. But in a learning experiment, for example, the probability of a correct response will usually increase with each correct response, while the probability of an incorrect response will decrease with each incorrect response, particularly if responses are corrected or reinforced. We are not merely iterating the truism that probabilities are relative to data. Rather we are urging that the data themselves depend for their import on the posited model of the experiment, the theoretical lens through which the outcomes are viewed and interpreted.

This point is often met by claiming the model itself must have been obtained empirically via the induction rule. One falls back, presumably, on higher-order inductive generalizations about light bulbs to infer that the probability of malfunctioning increases with use. But it is no less question-begging there than elsewhere to suppose that the background knowledge in point must have been gleaned by enumerative induction. The assumptions of standard probability models, like the normal or the binomial, describe idealized experiments which actual experiments are *designed* to approximate. It is only in an esoteric sense that models may be said to 'derive from experience'. The choice of a model is supported, rather, by mathematical analysis based on simplifying assumptions, as when one argues that a trait which depends on a large number of independently distributed dichotomous random variables is normally distributed. Galton boards are physical embodiments of this bit of abstract argument and were intended as such. I see no reason to exclude the possibility that an intelligent being could arrive at the correct normal curve for the scattering of balls in a Galton board by 'pure reason'. Hume apparently thought that if one chases mediating inductions back far enough, one will finally arrive at pure inductions supported directly by their instances – whence his curious discussion of the inductive inferences of a new-born babe. Certainly one characteristic that separates the brand of probabilism I espouse from inductivist positions is its skepticism of 'theoretically barren contexts'.

In any event, it is really irrelevant how we come by our probability models. For the problem is: given such background knowledge, how do we process it in making inductive inferences? How does a probability model combine with frequency counts to issue in a predictive probability? That the induction rule must be modified to take background knowledge into account is crashingly obvious. But it is far less obvious how the rule is to be modified. To invoke 'higher-order inductions' in lieu of mathematical analysis is to engage in the merest hand-waving. For the problem is to spell out how higher-order inductions mediate lower-order inductions.

Enter the inductivist's last stand. Granted, as we claim, that probability models determine the import of data, isn't it yet the case that when we base a conclusion on given experimental findings, we tacitly appeal to the induction rule in supposing that the experiment would issue in the same results were it replicated under similar conditions?

Predictive probabilities of future trial outcomes, as we have seen, are deductive consequences of the model and parameter estimates; nothing like the induction rule is presupposed. Perhaps the real locus of the objection is that the assumption of the same model in a replicated experiment rests on the induction rule. But here, I think, the inductivist has got the cart before the horse. We do not believe the same model will fit the current experiment because it has fit 'similar' experiments; rather, we classify experiments as similar to just the extent that they conform to the description of the same idealized experiment, that is, to the extent that they are susceptible of the same theoretical analysis.

3. UNIFORMITY AND DISCOVERY

The first line of defense against the charge that the induction rule fails to take cognizance of the mediating role of background theory is to insist that this knowledge must itself have been gleaned inductively via the rule. We saw that this contention is irrelevant, for even if it could be sustained, 'rich' contexts would have to be dealt with. But in this section we propose to scrutinize more fully the heuristic status of the rule, the claim that it alone enables us to uncover lawful regularity.

Given a table of data — mapping, e.g., the number of primes less than x against x, or the sum of the first n cubes against n — the induction rule not only fails to determine which of many discernible patterns to extrapolate, it does not even serve to *detect* pattern. It is absurd to depict the discovery of, say, the Prime Number Theorem or the formula for the sum of the first n cubes, as the result of extrapolating observed frequencies or concatenations of events. The whole difficulty in such problems is to find a simple formula that fits the instances already examined. Once the relationship has been found, its extrapolation is virtually automatic. The point is that the discovery of a law and its extrapolation are two separate problems. At best, the induction rule is pertinent to the second of these.

The deterministic or indeterministic character of the process under observation has no bearing at all on this point. Recursive sequences, like arithmetic progressions, are, I would suppose, deterministic paradigms. Yet, given the first n terms of such a sequence, e.g. 4, 12, 29, 57, 98, . . . , the problem of finding the recursive equations (or a formula for the general term) is no less acute for knowing such equations exist. It is not even assured that the correct equations will become more apparent as the number of terms is increased. A similar point applies to Reichenbach's discussion of the straight rule: knowing that sample proportions converge in the limit to the population proportion is no guarantee that the former will provide ever better estimates of the latter. We can only infer that this is true beyond *some* point. But that is no help unless we know *which* point. Similarly, if we knew that invalidity of first order formulae always shows up in a finite domain, that would not issue in a decision procedure unless we could set upper bounds! Nor would knowledge that at most finitely many causes govern an effect help us to isolate the 'active' cause. In any case, Reichenbach's argument for the straight rule founders on the fact that infinitely many other estimation rules (indeed, all those of Carnap's λ-continuum) are asymptotically indistinguishable from it.

There is no upper limit to the complexity of patterns discernable in data. Recording the ages at death of the first five poets in a biographical dictionary as 48, 76, 84, 48, 45 Peirce (1883) observes that any of the following relationships might be extrapolated: (i) the first digit raised to the power of the second is congruent to one modulo three, (ii) the difference in the first two digits is congruent to one modulo three, or (iii) the sum of the prime factors (including one) of each age is divisible by three. No one of these patterns is suggested by the induction rule, nor once recognized, does the rule accord any one of them a preferred status. To put this point in a somewhat more general form: a gambling system for a sequence of dichotomous trials might succeed by exploiting dependencies of any order on preceeding trial outcomes, drifts in the probability of success, or dependencies of the trial outcome on the place number of the trial. The induction rule is simply insensitive to the differences between these kinds of patterning.

The endless possibilities of superimposing pattern upon pattern render suspect the belief that any meaning can be attached to the uniformity of nature *sans phrase*. How could one ever distinguish a breakdown of such global uniformity from an additional dimension of complexity in the overall mosaic? How could one ever distinguish 'changes in the course of nature' from failure of particular laws? (Isn't the former but a metaphor for the latter?) Attempting to distinguish the two is not unlike attempting to distinguish absolute from relative motion.

Even if nature were uniform in some non-empty sense, that would have no tendency to buttress enumerative inductions. This is already clear from the example of recursive sequences. Hume's discussion proceeds as though the scientist discovers actual laws of nature which hold so long as the course of nature remains uniform. But this conception is already seriously inadequate, for if it were correct, then the numbers of instances already examined would have no bearing on the confidence one should place in an induction. All laws would be equally at the mercy of aberrant 'changes in the course of nature'. Reflection upon the essential vacuity of the principle of uniformity should foster appreciation of the kind of error committed in supposing that inductive inferences can be made in abstraction from the data generating process. It is not uniformity per se that inductive reasoning exploits, for there is no such animal, but rather the specific kind of uniformity embodied in the process under study.

4. CONDITIONALIZATION

There is no sound principle of induction as traditionally understood. Induction, or learning from experience, is just the process of revising probability assessments in the light of additional information. This is done by Bayes' rule:

(3.1) $P(H/x) = P(x/H)P(H)/P(x)$,

where H is an hypothesis and x an experimental outcome or datum. The problem of justifying induction is thus, in the first place, the problem of justifying conditionalization or Bayes' rule. Many such justifications have been given[7], but I want to sketch a particularly intuitive and compelling one here.

EXAMPLE 1. We are given an urn containing three balls, each of them either black or white. Label the four possible compositions 0, 1, 2, 3 indicating the number of black balls, and let H_i be the hypothesis that composition i obtains, $i = 0, 1, 2, 3$. Initially, let us suppose, the hypotheses are equiprobable. The contents of the urn are examined by a stooge who reports only that there is at least one black ball, thereby excluding H_0. What new probabilities should be assigned our hypotheses in the light of this information E?

Well, clearly, $P(H_0/E) = 0$, since E excludes H_0. While E excludes H_0, it does no more than this, and so has no bearing on the relative probabilities of the three non-excluded hypotheses. Since they were equiprobable prior to receipt of E, they should remain equiprobable. However, probabilities must sum to one, and so $P(H_i/E) = 1/3$, $i = 1, 2, 3$.

Essentially the same reasoning shows that, where initial probabilities are unequal, an outcome E which logically excludes H_0 but no more, should make the probability of H_0 zero and leave the relative probabilities of the non-excluded hypotheses unchanged. Thus, there is a constant of proportionality, c, such that $P(H_i/E) = cP(H_i)$, $i = 1, 2, 3$, and determined from the requirement that the new probabilities must continue to sum to one (i.e., c is a normalization constant). Solving $cP(H_1) + cP(H_2) + cP(H_3) = 1$ for c gives $c = 1/(P(H_1) + P(H_2) + P(H_3)) = 1/(1 - P(H_0))$. To say E logically excludes H_0 *and no more*, is to identify E with the denial of H_0, in which case our equation for c becomes $c = 1/P(E)$. Then the equation $P(H_i/E) = cP(H_i)$ reduces to $P(H_i/E) = P(H_i)/P(E)$, the special case $P(E/H_i) = 1$ of Bayes' rule. The probability $P(E/H_i)$ is conditional on an hypothesis; it is not 'conditional' in the sense of 'revised'. And these probabilities *conditional on an hypothesis* are easily computed for our problem. Given that H_i is true, for $i = 1, 2$, or 3, the probability is one that the contents of the urn include at least one black ball. I.e., $P(E/H_i)$ is indeed equal to one.

So much for the case of 'noiseless information' (Chapter 1). Suppose, next, that instead of sampling the entire urn, a single ball is drawn at random and found to be black (outcome B). Again H_0 is logically excluded, but this time, the more specific sample description B bears differentially on the non-excluded hypotheses H_i, $i = 1, 2, 3$. In fact, $P(B/H_i) = i/3$, $i = 0, 1, 2, 3$, and it is hard to resist the temptation to conclude that the revised probabilities $P(H_i/B)$ should be proportional to the *likelihoods* $P(B/H_i)$. For example, it is twice as 'likely' that a black ball be drawn from composition 2 as from composition 1, and so, if the initial distribution is uniform, as before, the revised probability of H_2 should be twice that of H_1.

This reasoning is certainly not without intuitive force. Can we give it a deeper foundation? To this end, imagine that some known chance mechanism determined the initial probabilities. Given that one of the balls thus placed in the urn was black, we might compute the new probabilities of the H_i by re-applying the same chance mechanism to place two additional balls in the urn. Any sound rule for revising initial probabilities would have to agree with the results of this calculation.

Assume, for example, that the composition of our urn was determined by drawing three balls at random (without replacement) from another urn containing 6 black balls and 4 white balls. The prior probabilities are given in Table I.

TABLE I

H_i	H_0	H_1	H_2	H_3
$P(H_i)$	1/30	9/30	15/30	5/30
$P(H_i/B)$	0	6/36	20/36	10/36

The revised probabilities $P(H_i/B)$ of the last row were obtained by re-applying the chance mechanism in question. Suppose, first, that the black ball sampled was the first of the three placed in the urn. The new probability of H_1, for instance, is then the probability that no blacks will be drawn in sampling two balls without replacement from an urn containing 5 black and 4 white balls, or $\binom{5}{0}\binom{4}{2}\Big/\binom{9}{2}$ = 6/36. The other entries in the last row of Table I were similarly obtained. Notice, next, it makes no difference whether the black ball sampled from our urn was placed there first, second, or third by the chance mechanism. If it was placed there second or third, we can regard it as 'destined' for placement, and proceed as if it had been placed first, obtaining again the last row of Table I. Finally, as you can quickly verify, the entries in the last row of Table I agree with the posterior probabilities for this problem computed by Bayes' rule, using the partition property $P(B) = P(B/H_0)P(H_0) + \cdots + P(B/H_3)P(H_3) = 0.6$. Quite obviously, the composition of the 'super' urn is also immaterial; the same agreement between the two methods of computing the revised probabilities obtains.

Moreover, this agreement will be preserved whatever chance mechanism is used to determine the composition of our urn. To take another case, an (ideally) fair coin is tossed three times. Each time a head turns up, a black ball is placed in the urn; each time a tail turns up, a white ball is placed in the urn. The initial probabilities are then as given in Table II.

The probabilities $P(H_i)$ are obtained simply by dividing the number of outcomes of the three flips which produce the given composition by the total number of outcomes (eight), since all eight outcomes are equiprobable. One of the balls having been placed, we consider the outcome of two flips. The

TABLE II

H_i	H_0	H_1	H_2	H_3
$P(H_i)$	1/8	3/8	3/8	1/8
$P(H_i/B)$	0	1/4	1/2	1/4

revised probability of H_i is then the probability that $i - 1$ of two flips show heads, $i = 1$, 2, 3, while, of course, $P(H_0/B) = 0$ as before. These are the probabilities shown in the last row of Table II. Again they could have been obtained by conditionalization, i.e., by multiplying the likelihoods 0, 1/3, 2/3, 1 by the prior probabilities (first row) and re-normalizing.

To see what is going on, I list all eight possible outcomes of the three tosses. Again, it makes no difference to the analysis whether the black ball sampled was the first, second, or third ball placed in the urn by the chance device, so I assume it was placed first. The outcomes marked with an asterisk are those which satisfy H_1. Given our assumption that the black ball was placed first, $B = \{HHH, HHT, HTH, HTT\}$. Write $P_B(H_i)$ for the revised probability of H_i computed by the second method (re-applying the chance device).

$$
\left.
\begin{array}{l}
HHH \\
HHT \\
HTH \\
*HTT
\end{array}
\right\} B
$$

$$
\begin{array}{l}
THH \\
*THT \\
*TTH \\
TTT
\end{array}
$$

The $P_B(H_i)$ are then found by toting the probabilities $P(x/B)$ of elements x of B which realize H_i. By our earlier argument (for the noiseless case), these probabilities $P(x/B)$ are proportional to their initial values $P(x)$ and sum to one: $P(x/B) = P(x)/P(B)$. (Thus, in our example, the elements of B are equiprobable.) E.g., for H_1, HTT is the one element of B which satisfies the hypothesis, whence $P_B(H_1) = 1/4$. But HTT is the intersection of B with H_1, and so $P_B(H_1) = P(HTT/B) = P(HTT)/P(B) = P(H_1, B)/P(B)$. Finally, $P(H_1, B) = P(H_1/B)P(B)$, since the proportion of outcomes which realize H_1 and B is the proportion of B-outcomes which realize H_1 multiplied by the proportion of B-outcomes. (It is easily seen, moreover, that this last identity continues to hold when the outcomes are unequally probable.) We conclude:

$$(3.2) \qquad P_B(H_1) = P(H_1/B) ,$$

and the same argument shows that the two methods of revising initial probabilities agree on the other H_i as well. Even though explained with reference to a particular chance mechanism, our argument is completely general. Its effect is to reduce the case of noisy information (outcomes which bear differentially on non-excluded hypotheses) to the case of noiseless information considered above.

To be sure, our discussion has been limited to the case where a chance device determines initial probabilities. The strategy of our justification applies, however, where knowledge of any sort determines the prior probabilities. If an item of information logically compatible with that knowledge is added, we have only to re-compute the probabilities of the considered hypotheses for that augmented stock of information. In every case, we will obtain the same result by conditionalizing on that new item of information, starting from the initially given probabilities. Where a new item of information conflicts with previously given information or shows it to be erroneous, we throw away that defective information and compute probabilities afresh. It is in this sense that prior probabilities may be said to be 'corrigible' (as opposed to modifiable by conditionalization); in such cases, conditionalization is superseded.

Before leaving this topic, I note the possibility, emphasized by Richard Jeffrey, of conditionalizing on uncertain outcomes. Suppose, in our example, where E is equivalent to the denial of H_0, that the stooge who reports E is only 90% reliable. The revised probabilities P^* are then obtained by conditionalizing iteratively both on E and on this additional information:
$$P^*(H_i) = P(H_i/E)P(E) + P(H_i/\overline{E})P(\overline{E}) = 0.9P(H_i/E) + 0.1P(H_i/\overline{E}), \ i = 0, 1, 2, 3.$$

5. THE MAXIMUM ENTROPY RULE

Our justification of Bayes' rule rests on a form of the Principle of Insufficient Reason. Traditionally, this principle has been more often used to ground prior probability distributions than their mode of revision. But, allegedly, it leads to contradictions – the Bertrand Paradox, for one. These putative contradictions led many to abandon the Bayesian approach, or rather, the attempt to assign probabilities to hypotheses, altogether. But even among those who adhere to this program, there is a general abandonment of any pretense to the objective representation of partial knowledge by a probability distribution. Instead, 'subjective Bayesians' argue that two rational men faced with the same data may rationally diverge in their probability assignments to the same alternatives. They are bound only by the requirement that their assignments do not violate the axioms of probability.[8]

On the other hand, there has been a dissenting modern Bayesian tradition – a sort of fundamentalism – which does maintain that states of knowledge can sometimes uniquely determine a probability distribution. Though popularly associated with names like Keynes, Jeffreys, and Carnap, this school of thought has in fact received its most powerful expression and most sophisticated defense in the writings of the physicist, E. T. Jaynes. It would be well to quote here from Jaynes' fundamental paper, 'Prior Probabilities' (Jaynes, 1968) at some length:

Surely, the most elementary requirement of consistency demands that two persons with the same relevant prior information should assign the same prior probabilities. Personalistic doctrine makes no attempt to meet this requirement, but instead attacks it as representing a naive 'necessary' view of probability, and even proclaims as one of its fundamental tenets ... that we are free to violate it without being unreasonable. Consequently, the theory of personalistic probability has come under severe criticism from orthodox statisticians who have seen in it an attempt to destroy the 'objectivity' of statistical inference by injecting the user's personal opinions into it.

Of course, no one denies that personal opinions are entitled to consideration and respect if they are based on factual evidence .... Nevertheless, the author must agree with the conclusions of orthodox statisticians, that the notion of personalistic probability belongs to the field of psychology and has no place in applied statistics. Or, to state this more constructively, objectivity requires that a statistical analysis should make use, not of anybody's personal opinions, but rather the specific factual data on which those opinions are based. ...

Evidently, then, we need to find a middle ground between the orthodox and personalistic approaches, which will give us just one prior distribution for a given state of prior knowledge. Historically, orthodox rejection of Bayesian methods was not based at first on any ideological dogma about the 'meaning of probability' and certainly not on any failure to recognize the importance of prior information .... The really fundamental objection ... was the lack of any principle by which the prior probabilities could be made objective ... Bayesian methods, for all their advantages, will not be entirely satisfactory until we face the problem squarely and show how this requirement can be met.

From an objectivist Bayesian standpoint, then, prior probabilities are as objective, qua mathematical representation of prior information, as likelihoods. The latter, after all, are but consequences of the assumptions embodied in a probability model, and these assumptions need not be wholly realistic for the model to serve. For, while expecting the assumptions to oversimplify the phenomena, the direction of departure may be unknown, the abandonment of an assumption may lead to intractable mathematics, or slight departures from an assumption may be shown to have inappreciable affects on the accuracy of the predictions generated.

Prior probabilities are often based on assumptions with precisely these characteristics. The celebrated problem of calculating the probability that, in a group of n persons haphazardly collected, at least two will have the same birthday, is a good illustration of this. We assume that the birthdays drawn by different persons represent independent random selections from the 365 possibilities. Thus, while there may well be some tendency of birthdays to cluster, nothing being known of what particular form the clustering takes, it is most reasonable to assume no clustering occurs. Given these assumptions, the complementary probability that no two of the n persons have the same birthday is found by dividing the number of ways n persons can be randomly assigned to 365 possible dates *without repetition* by the total number of ways of assigning each of n persons one of 365 dates: Viz., $(365)_n/365^n$, where $(365)_n = 365 \cdot 364 \ldots (365 - n + 1)$ is the nth falling factorial of 365. To assume clustering would enormously complicate this simple calculation, and, unless clustering were quite marked, it would not appreciably affect the result.

What has been found objectionable is the suggestion that, lacking any grounds for thinking that there is clustering of birthdays (or a 'mating season' in a given human population), or lacking knowledge of the particular form clustering takes, we are justified in assuming the absence of clustering. It seems a case, as R. A. Fisher put it, of assuming knowledge of that of which in fact we are ignorant. In practice, though, scientists do this all the time, and it is curious that many who do it in arriving at a theory of some phenomenon object to such reasoning when it underpins a prior distribution. Steady-state cosmologists, to take one of myriad instances, start off by assuming the laws of physics are the same in temporally and spatially remote regions of the universe. This, they urge, is surely the simplest assumption. But it is more than that. To assume that different laws obtained a billion years ago would be entirely arbitrary; it would be to import knowledge we do not in fact possess. Those who assumed Newton's gravitation law as a working hypothesis in first treating the motions of binary stars were making another 'conservative extrapolation'. They were going as little beyond existing knowledge as possible. (Newton himself laid down this principle as one of the 'rules of reasoning' in the *Scholium* of the *Principia*.) One must say what it is that makes an otherwise sound principle objectionable when applied to prior probabilities.

Not every problem of probability resolves naturally into a partition of equiprobable atoms or simple events. To uphold an objective Bayesian position, one must extend Laplace's form of the Principle of Insufficient Reason to a method of objectively representing any prior information. Taking our clue from Newton's Principle of Conservative Extrapolation, and following E. T. Jaynes, we arrive at the *maximum entropy rule*: *choose that prior which, among all those consistent with the prior information, maximizes uncertainty.* Where prior information assumes the form of constraints governing the relevant random variable, we maximize entropy subject to the given constraints. We illustrate this principle below. The other part of the problem facing objectivist Bayesians is to dispose of the alleged contradictions to which Laplace's special case of this principle leads. This involves considerations of invariance which will occupy us in the next section. Consider first the case where the prior constraints are mean values of one or more functions $f_i(X)$ of the discrete random variable X, which assumes possible values x_i with probability p_i, $i = 1, \ldots, n$. Our problem is to maximize $H = -\Sigma p_i \log p_i$ subject to the constraints $\Sigma p_i = 1$ and the m mean value constraints $\Sigma_i p_i f_k(X_i) = F_k$, $k = 1, \ldots, m$. The solution is:

$$(3.3) \qquad p_i = \frac{1}{Z(\lambda_1, \ldots, \lambda_m)} \exp[\lambda_1 f_1(X_i) + \cdots + \lambda_m f_m(X_i)] \,,$$

with

$$(3.4) \qquad Z(\lambda_1, \ldots, \lambda_m) = \sum_{i=1}^{n} \exp[\lambda_1 f_1(X_i) = \cdots + \lambda_m f_m(X_i)] \,,$$

with the λ_k determined from

$$(3.5) \qquad F_k = \frac{\partial}{\partial \lambda_k} \log Z(\lambda_1, \ldots, \lambda_m) \,.$$

I give the derivation of (3.3)–(3.5) for the case $m = 1$ of just one mean value constraint; the general case follows by an utterly straightforward extension of the proof. By Lagrange's method, we must solve:

$$0 = \frac{\partial}{\partial p_i} [H + \mu(\Sigma p_i - 1) + \lambda(\Sigma p_i f(X_i) - F) \,,$$

$$= -(\log p_i + 1) + \mu + \lambda f(X_i) \,,$$

for p_i, giving

$$\log p_i = \lambda f(X_i) + \mu - 1 \,,$$

or

$$p_i = C \exp(\lambda f(X_i)) \,,$$

where C is a normalization constant. The constraint $\Sigma p_i = 1$ then gives

$C = Z(\lambda)$, given by (3.4), with $m = 1$. For (3.5):

$$\frac{\partial}{\partial \lambda} \log Z(\lambda) = \frac{1}{Z(\lambda)} \frac{\partial}{\partial \lambda} [e^{\lambda f(X_1)} + \cdots + e^{\lambda f(X_m)}],$$

$$= f(X_1) \frac{e^{\lambda f(X_1)}}{Z(\lambda)} + \cdots + f(X_n) \frac{e^{\lambda f(X_n)}}{Z(\lambda)},$$

$$= \Sigma p_i f(X_i),$$

$$= F.$$

EXAMPLE 1. Given a discrete random variable X which assumes non-negative integer values $0, 1, 2, \ldots$, and $E(X) = 5$, to find the maximum entropy distribution of X.

Solution. Here $f(X_k) = X_k = k$, whence,

$$p_k = \frac{e^{\lambda k}}{\Sigma e^{\lambda k}} = e^{\lambda k}(1 - e^{\lambda}),$$

upon summing the geometric series $\Sigma e^{\lambda k}$. By (3.5):

$$5 = \left| \frac{\partial}{\partial \lambda} \log(1 - e^{\lambda}) \right| = e^{\lambda}/(1 - e^{\lambda}),$$

whence

$$e^{\lambda} = 5/6 \quad \text{and}$$

(3.6) $$p_k = P(X = k) = \left(\frac{5}{6}\right)^k \left(\frac{1}{6}\right), \qquad k = 0, 1, 2, \ldots .$$

We recognize (3.6) as a case of the *geometric distribution* (a special case of the negative binomial distribution):

(3.7) $$p_k = p^k q, q = 1 - p.$$

This distribution gives the probability that k trials will precede the first 'failure' in a sequence of Bernoulli trials. Similarly, if X is a continuous non-negative random variable, its maximum entropy distribution is found to be the *exponential distribution*

(3.8) $$P(X = x) = \exp(-x/\mu)/\mu$$

the continuous analogue of the geometric distribution and a special case of the gamma distribution, the continuous analogue of the negative binomial.

Combined with invariance, the maximum entropy rule is an extremely powerful tool, as this example already suggests, not only for obtaining prior distributions, but probability models of empirical phenomena as well.[9] Much less is needed than is used in the standard derivations of the familiar probability models of statistics, like the normal, gamma, Poisson, etc. By comparison with the above derivation based on maximum entropy, these standard derivations are both ad hoc and overly elaborate.

EXAMPLE 2. Computationally, it is harder to find the maximum entropy distribution of a finitely many-valued discrete variate X. E.g., let K be the number of spots turned up when a single six-sided die is cast. If $E(K) = 3.5$, the maximum entropy distribution of K is uniform (as we should expect). If $E(K) = 4$, on the other hand, we expect a distribution gently skewed to the right. Using (3.3) and (3.4) with $f(x_k) = k$, $k = 1, \ldots, 6$, we obtain

$$p_k = e^{\lambda k}/Z(\lambda) \,,$$

with $Z(\lambda) = e^\lambda + e^{2\lambda} + \cdots + e^{6\lambda}$, and λ determined by $(\partial/\partial\lambda) \log Z(\lambda) = 4$. The latter equation reduces to: $2u^5 + u^4 - u^2 - 2u - 3 = 0$, with $u = e^\lambda$. Solving numerically, $u \doteq 1.190804264$, whence $\lambda \doteq 0.1746289309$ and $Z(\lambda) \doteq 11.55388764$. The values of the p_k are listed in Table III.

TABLE III

k	1	2	3	4	5	6
p_k	0.103	0.123	0.146	0.174	0.207	0.247

Suppose the die is cast and we are told only that an ace did not turn up. Bayes' rule yields the five non-zero probabilities $q_k \propto p_k$, $k = 2, \ldots, 6$, of Table IV. Let Y be the random variable which assumes possible values $2, \ldots, 6$ with the probabilities of Table IV. Then $E(Y) = 4.344$. To compute the maximum entropy distribution of Y, we must solve

$$\frac{1}{Z(\lambda)} 2e^{2\lambda} + \cdots + 6e^{6\lambda} = 4.344 \,,$$

TABLE IV

k	2	3	4	5	6
q_k	0.137	0.163	0.194	0.321	0.275

for λ, with $Z(\lambda) = e^{2\lambda} + \cdots + e^{6\lambda}$. Solving numerically, as before, we find that the maximum entropy distribution of Y agrees with that of Table IV.

EXAMPLE 3. Consider again the variate K of Example 2. This time we are given the qualitative constraint $p_2 = p_4 + p_5$. To find the maximum entropy distribution subject to this constraint, we maximize $H = -p_1 \log p_1 - 2p_2 \log p_2 + p_2 \log 2 - p_3 \log p_3 p_6 \log p_6$, subject to $p_1 + 2p_2 + p_3 + p_6 = 1$, since, clearly, $p_4 = p_5$ in the maximum entropy distribution in question. Here we obtain:

$$p_k = \begin{cases} C & \text{for} \quad k = 1, 3, 6 . \\ \sqrt{2}C & \text{for} \quad k = 2 . \end{cases}$$

TABLE V

k	1	2	3	4	5	6
p_k	0.172	0.242	0.172	0.121	0.121	0.172

Solving $1 = 2\sqrt{2}C + 3C$ for C gives $C = 0.1715728753$, and the p_k, with $p_4 = p_5 = p_2/2$ are given in Table V. Suppose now that we are given $K \neq 1$. This adds the constraint $p_1 = 0$ and is consistent with the old constraint $p_2 = p_4 + p_5$. (Given $K \neq 1$, our problem is to guess which of the non-excluded values of K did occur.) With the new constraint added, we again obtain $p_k = C$ for $k = 3, 6$ and $p_2 = \sqrt{2}C$, whereupon, solving $\sqrt{2}C + 2C = 1$ for C, we get $C = 0.2071067812$, and the following values of the p_k:

k	1	2	3	4	5	6
p_k	0	0.293	0.207	0.1465	0.1465	0.207

the same values obtained by conditionalization. Hence, given qualitative constraints and information which excludes some possibilities but is consistent with those constraints, *the new maximum entropy distribution for the enlarged set of constraints is the same distribution obtained from the old one by conditionalization.*

I have deliberately emphasized the correspondence between conditionalization and the maximum entropy rule, but the latter is clearly, in our account, the more fundamental. We feel justified in imposing on any rule for updating probability distributions in the light of sample data the constraint that it lead from maximum entropy distributions to maximum entropy distributions (with the proviso that the sample data be consistent with the prior constraints).

Of course, not all pertinent data comes to us in a form to which Bayes' rule can be applied. We might be given information about the mean of a random variable, then additional information about its variance or higher moments. No item in this expanding sequence can be conditionalized upon. Instead, we obtain ever stronger constraints which, if consistent, can serve as inputs in applying the maximum entropy rule.[10] Good (1975) has introduced the term 'evolving probabilities' to cover cases of this sort.

Statistical mechanics provides further illustration of these evolving probabilities. We are given a physical system composed of n identical particles (e.g., photons, protons, electrons, mesons), each one of which can be in any of S microstates (which might be energy levels). The set of S-tuples $(n_1, \ldots, n_S)$ whose jth coordinate, n_j, gives the number of particles in the jth microstate, $j = 1, \cdots, S$, is called *phase space*. The points (S-tuples) of this space are the *macrostates* of the system. The *equilibrium state* is the most probable macrostate. If the particles are physically distinguishable, each of the n^S microstates will have equal probability (Maxwell-Boltzmann statistics). But if the particles cannot be individuated, the $n!$ microstates arising from permutations of the n particles must be combined into a single alternative (what Carnap (1962) terms a 'structure description'), and each structure description will be assigned equal probability by the maximum entropy rule (Bose-Einstein statistics). Finally, we might be given the additional information that the particles satisfy Pauli's Exclusion Principle, that no two particles are simultaneously in the same microstate, and in that case, our principle directs us to assign equal probability to each of the $\binom{S}{n}$ assignments of the n particles to distinct microstates (Fermi-Dirac statistics).

For a long time it was thought necessary on a priori grounds that all small particles have Maxwell-Boltzmann statistics. So far, no particles for which this is the appropriate model have been found. The moral usually drawn[11] is that one cannot discover the probability law of a physical system by a priori (e.g., Laplacean) reasoning. I should say, instead, that the misplaced faith in Maxwell-Boltzmann statistics was based, not on faulty reasoning or faulty principles of probability, but solely on the mistaken assumption that small particles are physically distinguishable. (It is no criticism of the maximum entropy rule that it leads to incorrect results when you feed it misinformation.) Once an erroneous assumption like this one is recognized, the probabilities based on the Maxwell-Boltzmann model must be corrected (by being rederived from a more adequate model), and not simply updated by Bayes' rule. Insofar as the prior constraints on which the maximum entropy rule operates are empirically obtained, the probability distribution yielded by the rule are empirically grounded. Given the correctness and exhaustiveness of those constraints, however, there may be reason to believe that maximum entropy distributions satisfying those constraints are 'physically necessary' in a certain sense. Let us consider an argument Jaynes (1968) propounds to this effect.

Given the correctness and exhaustiveness of the prior constraints, the maximum entropy distribution will be overwhelmingly the most likely to be observed experimentally, because it can be realized in by far the greatest number of ways.[12] The 'physical necessity' attaching to a maximum entropy distribution is therefore of a piece with that attaching to irreversibility in thermodynamics. (Indeed, as Jaynes has shown[13], the major probability distributions of classical statistical mechanics can be obtained, given mean values of the energy, by the maximum entropy rule, and, as he also points out, Gibbs himself (cf. Gibbs (1902), Chapter 11, esp. p. 143) appears to have anticipated Jaynes in this use of the rule.)

The special case of this correspondence with frequencies most relevant to statistical mechanics, viz., discrete random variables with mean value constraints, is treated in Jaynes (1968), Section IV. Thus, let X assume possible values $X_1, \ldots, X_n$ with (unknown) probabilities $p_1, \ldots, p_n$ (n could be infinite), and assume that the mean values of $M \leqslant n - 1$ functions f_i of X are given. The maximum entropy distribution is then given by (3.3)–(3.5) above. In M repetitions of the experiment, let m_i be the number of times outcome X_i occurs, so that

(1) $\Sigma m_i = M$.

Given the correctness of the mean values, we also have:

(2) $m_i f_k(X_i) = MF_k, k = 1, \ldots, m$,

where F_k is the mean value of f_k. The number of occasions on which the sample numbers $(m_1, \ldots, m_n)$ occur in M repetitions is given by the multinomial coefficient:

$$(3) \qquad W = \frac{M!}{m_1! \ldots m_n!} = \frac{M!}{(Mr_1) \ldots (Mr_n)!},$$

and so the set $(r_1, \ldots, r_n)$ of relative frequencies which can be realized most ways is that which maximizes (3), or any monotone transformation thereof, subject to the constraint (2). Consider now the monotonic function $M^{-1} \ln W$ of (3). As $M \to \infty$, Stirling's approximation gives

$$(4) \qquad M^{-1} \ln W \to - \Sigma r_i \ln r_i.$$

We are thus led to maximizing the entropy of the frequencies subject to m mean value constraints, precisely the problem of which the maximum entropy distribution (3.3)–(3.5) is the solution, with the r_i in place of p_i, and so the p_i are indeed the frequencies which can be realized in the greatest number of ways in a large number of repetitions of the experiment. Moreover, as Jaynes also shows, the maximum of W is enormously sharp; the maximum entropy distribution can be realized in overwhelmingly more ways than any other, the discrepancy increasing with the number of repetitions of the experiment.

When maximum entropy distributions fail to fit empirical data, we may therefore be sure that additional constraints (or different constraints) are operative. This presupposes, of course, that the n-tuples $(m_1, \ldots, m_n)$ which can be realised in most ways are the most probable, and this is close to assuming that all such n-tuples are equiprobable in the absence of positive indications to the contrary. If Jaynes' result were offered as justification of the entropy rule, then, that justification would be essentially circular. Still, those who do find the rule compelling will draw the above inference. E.g., when the maximum entropy distribution Gibbs derived (his 'canonical ensemble') failed to predict heat capacities and equations of state correctly, the existence of an additional constraint was indicated, and it turned out to be the discreteness of the energy levels. In this case, failure of a maximum entropy distribution provided one of the early clues leading to the quantum theory.

How general is the method of maximum entropy? Ideally speaking, it applies wherever we have what Jaynes calls 'testable information', viz., information whose consistency with a probability distribution can be ascertained. Needless to say, what can be ascertained in practice is a function of the knowledge and computational aids available at the time. And, of course, the information in question must be consistent, and we lack an algorithm for attesting consistency. Such limitations are inherent and hardly count against the method as such.

More serious is the fact that maximum entropy distributions do not always exist. If, for instance, we weaken the definite constraint that $E(X) = m$ in Example 1 to require only that $E(X) < \infty$, then no maximum entropy distribution exists.[14] In many such cases, however, a natural extension of the method recommends itself. In the case before us, we could treat m itself as a random variable about which nothing is known (beyond $0 < m < \infty$), and then average the maximum entropy distributions

$$p_k = \frac{1}{m} \left(\frac{m}{m+1} \right)^k$$

satisfying $E(X) = m$ against the density $f(m)$ of m, to obtain

$$p_k = \int_0^\infty \frac{1}{m} \left(\frac{m}{m+1} \right)^k f(m) \mathrm{d}m, \quad k = 0, 1, 2, \dots .$$

Further extensions of the method are contemplated in the next section.

6. INVARIANCE

Before the maximum entropy rule can be considered a complete solution of the problem of representing states of partial knowledge, it must be shown how to treat the case of no prior knowledge or complete ignorance. For one thing, as Jaynes has shown, the solution of this problem is presupposed in extending the maximum entropy rule from discrete to continuous distributions.[15]

Bayes and Laplace proposed uniform priors to represent total ignorance. The difficulty is that the uniform prior is not invariant — transformation of θ to another parameter, like θ^{-1} or $\log \theta$ determines a non-uniform distribution for it. But if we are ignorant of θ, or so the argument runs, then, equally, we are ignorant of $T(\theta)$, and so $T(\theta)$, too, should be uniformly distributed. But in general, the distribution of $T(\theta)$ determined from the uniform distribution of θ will be non-uniform, and so we have a contradiction: the distribution of $T(\theta)$ is both uniform and non-uniform. Contradictions of this sort based on the alleged 'arbitrariness of parametrization' have long been considered the rock upon which the whole Bayesian approach founders. The 'paradoxes of geometric probability' proposed by J. Bertrand in the nineteenth century are religiously cited in this connection.

The retreat to subjectivism (or 'personalism') by modern Bayesians was no doubt inspired in part by the conviction that these problems could not be overcome. In a context where substantial data is lacking, a uniform prior for θ might seem no more defensible than a uniform prior for, say, $\cos \theta$. It has only been quite recently that Jaynes, building on earlier suggestions of Sir Harold Jeffreys, has evolved a more adequate Bayesian response. The thrust of Jaynes' work is to use the invariance requirements implicit in the very statement of a problem to determine the appropriate prior distribution. The needed invariances, however, are not obtained by looking at parameter transformations *per se*, but at transformations of the problem itself into equivalent form. Given the statement of the problem, it may, for example, be indifferent in what scale units the data are expressed. Such 'indifference between problems' determines what parameter transformations are admissible – not the other way around. In some cases, as we will shortly see, the resulting invariance requirements uniquely determine a prior distribution (the same one, of course, for each 'equivalent' reformulation of the problem). Where no unique distribution is determined, the principle of maximum entropy may be invoked. Finally, the problem may be 'overdetermined' in the sense that no probability distribution satisfies all the invariance requirements. All that one can conclude then is that the requirements are inconsistent and the problem ill-posed. Before showing how the invariance method can be used to resolve the Bertrand paradox, I illustrate its application in a number of other cases.

EXAMPLE 4 (Poisson rates). In a Poisson process, the probability that n events will occur in a time interval t is:

(1) $\qquad p(n/\lambda, t) = \exp(-\lambda t)(\lambda t)^n / n!$.

Suppose we are initially ignorant of the rate parameter, λ, apart from its having physical dimensions $(\text{seconds})^{-1}$. Thus, we are in ignorance of the absolute time scale of the process. Now Jaynes (1968) argues as follows.

Assume that the clocks of Mr X and Mr X' run at different rates, so that their measurements of a given interval of time are related by:

(2) $\qquad t = qt'$.

Since they are observing the same physical experiment, their rate constants, λ and λ', must be related by $\lambda't' = \lambda t$, whence $\lambda'qt' = \lambda qt$, or

(3) $\lambda' = q\lambda$.

Their respective prior distributions are:

(4) $p(d\lambda/X) = f(\lambda)d\lambda$

and

(5) $p(d\lambda'/X') = g(\lambda')d\lambda'$.

For (4) and (5) to have the same content requires the *transformation condition*

(6) $f(\lambda) = qg(\lambda')$,

since $d\lambda' = q\ d\lambda$. (I.e., q^{-1} is the Jacobian of the transformation (3).) Now the probability distribution of λ should not depend on what clocks are being used (the *invariance requirement*), and so f, and g should be identical functions.

(7) $f(\lambda) = g(\lambda)$.

Combining the transformation condition (6) with the invariance requirement (7) yields the functional equation

(8) $f(\lambda) = qf(q\lambda)$

whose solution (up to a proportionality constant) is

(9) $f(\lambda) = 1/\lambda$.

The density (9) does not integrate to one; it is an *improper* prior. It can be combined with a likelihood function, however, to yield a proper posterior distribution of λ, viz.:

(10) $p(d\lambda/n) \propto p(n/\lambda, t)\lambda^{-1} d\lambda \propto e^{-\lambda t} \dfrac{(\lambda t)^{n-1}}{(n-1)!} (t\ d\lambda)$,

upon absorbing $1/n$ into the constant of proportionality. (10) is a proper density (integrating to one). Generally speaking, improper priors should be regarded as convenient approximations to a proper prior which is flat in the region where the likelihood is substantial, but which vanishes rapidly enough outside of that region to be integrable. Indeed, any proper prior with these properties will yield posterior distributions of the parameter scarcely distinguishable from the posterior distributions based on an improper prior. (Precisely formulated, L. J. Savage referred to this result as 'the principle of stable estimation'.)[16]

Suppose next that λ is constrained to lie in a finite interval $[A,B]$, but that nothing else is known. The same invariance argument leads to what is now the proper prior (9), with normalization constant equal to $\log(B - A)$. (For convenience, the base of the logarithms can be chosen so that $\log(B - A) = 1$.) Hence, the probability that λ lies in a subinterval $[C, D]$ of $[A, B]$ is $\log(D - C)/\log(B - A)$, and not the quantity $(D - C)/(B - A)$ associated with the uniform density over the interval $[A, B]$. The (now proper) prior (9) remains invariant under change of scale. I will refer to this prior throughout as the *log-uniform distribution*. (N.B., X has the log-uniform distribution in $[A, B]$ if $\log X$ has the uniform distribution in $[\log A, \log B]$.)

The invariance argument applied to scale-invariant problems has obvious application to the von Kries paradox (Keynes, 1921, pp. 45f.).

Let V be the specific volume of a certain substance, and suppose that V is known to lie in the closed interval $[1,3]$. Classically, V would be assigned a uniform distribution. Now $v = 1/V$ is the specific density and lies in the interval $[1/3, 1]$. We are then faced with the usual contradiction, for the uniform distribution of V induces a non-uniform distribution of $1/V = v$. The way out, of course, is to recognize the scale invariance implicit in the very statement of the problem (nothing having been said about the particular scale employed). The requirement of scale invariance, we have seen, determines, not a uniform, but a log-uniform distribution of our parameter V, and this distribution is invariant under the parameter transformation $T(V) = 1/V$. Not any old transformation need be admissible, on the other hand, and to determine whether a transformation is admissible, we must, in general, first determine the prior of the given parameter from the group of transformations of the *problem* which, from the standpoint of our partial knowledge, constitute equivalent reformulations of the problem. And, notice, in the von Kries problem, it makes no difference whether we look first at specific volume or specific density. Arguments from scale invariance will also be used in our next two examples.

EXAMPLE 5 (location and scale parameters). Given a probability density h with a location parameter μ and a scale parameter σ, about which we have no prior knowledge, our problem is to estimate the parameters from a random sample. In the absence of prior knowledge it might be natural to *assume* that the parameters are independent. We would then arrive at a uniform distribution for the location parameter and a log-uniform distribution for the scale parameter (as in Example 4). The joint prior density would then be:

$$(3.9) \qquad f(\mu, \sigma) \propto 1/\sigma .$$

On the other hand, as Jaynes (1968) shows, (3.9) can be based on a simple invariance argument which does not assume independence of μ and σ. For, to characterize μ as a location parameter about which nothing is known is to suggest the joint distribution should be invariant under the translation

$$(1) \qquad \mu' = \mu + b, \ -\infty < b < \infty.$$

Further, since σ is a scale parameter about which nothing is known, the joint prior should be invariant under change of scale:

$$(2) \qquad \sigma' = a\sigma, 0 < a < \infty.$$

Finally, the values of the random variable in the two scales are related by:

$$(3) \qquad x' - \mu' = a(x - \mu) .$$

Invariance under the transformation $(x, \mu, \sigma) \to (x', \mu', \sigma')$ with x', μ', σ' given by (1)–(3) determines the joint density as (3.9).

Suppose, for example, that h is the normal density. In a random sample $(x_1, \ldots, x_n)$ of n, the densities multiply to give the likelihood:

$$p(x/\mu, \sigma) \propto \sigma^{-n} \exp\left[-\frac{1}{2\sigma^2} \Sigma(x_i - \mu)^2\right] .$$

But $\quad \Sigma(x_i - \mu)^2 = \Sigma(x_i - \bar{x} + \bar{x} - \mu)^2 = \Sigma(x_i - \bar{x})^2 + n(\bar{x} - \mu)^2, \quad$ since $\Sigma(x_i - \bar{x}) = 0$. Writing $S_{\bar{x}\bar{x}} = \Sigma(x_i - \bar{x})^2$, the posterior density based on a random sample of n and the informationless (improper) prior (3.9) is:

$$(3.10) \qquad p(\mu, \sigma/x) \propto \sigma^{-(n+1)} \exp(-\frac{1}{2\sigma^2} [S_{\bar{x}\bar{x}} + n(\mu - \bar{x})^2]) ,$$

Although μ and σ are independent according to (3.9), they are dependent according to (3.10), the degree of dependence, however, decreasing with n. For most practical purposes, then, one can work with the marginals for making inferences about either parameter separately. The marginal densities are:

$$(3.11) \qquad p(\mu/x) = stu(\mu/\bar{x}, \frac{x^2}{n}, n - 1), \quad s^2 = S_{\bar{x}\bar{x}}/(n - 1)$$

and

(3.12) $p(\sigma/x) = igam(\sigma/q, n - 1)$, $q^2 = S_{\bar{x}\bar{x}}/n$ is the sample variance.

(3.11) is Student's distribution with $n - 1$ degrees of freedom (df), and (3.12) is the inverse gamma distribution. Tables of stu may be found in almost any statistics text, while tables of $igam$ are given in Schmitt (1969).

For the case of known σ, the posterior distribution of μ is itself normal about the sample mean with variance σ^2/n. This distribution is slightly more peaked (and smaller in the tails) than stu, in which s^2 replaces σ^2. As intuition suggests, stu approaches the normal distribution in question as $n \to \infty$.

I would also point out that (3.11) and (3.12) are only defined for $n > 1$. Thus, no inferences about either the mean or the variance can be based on a single observation when both these parameters are unknown. Given a single observation, $s^2 = q^2 = 0$, which is further indicative of degeneracy. How could anything be learned about the variance from a single observation when the mean is unknown? When μ is known, by contrast, one can make inferences about σ given only one observation. E.g., if $\mu = 0$ and $x_1 = 1$, the ML estimate of σ^2 is found by determining the normal curve, centered at zero, which has the highest ordinate at $x_1 = 1$. From (3.10), with $n = 1, \mu = 0$, and $x = 1$, the log-likelihood is $\propto - \ln \sigma - 1/(2\sigma^2)$. Differentiating and setting the result equal to zero, we obtain $\delta^2 = 1$. The mode of the posterior distribution based on the Jeffreys prior (3.9), found by differentiating $\ln p(\sigma^2/x) \propto - 2\ln \sigma - 1/(2\sigma^2)$, is $\sigma^2 = 0.5$. The discrepancy between the two estimates diminishes, of course, with increasing sample size. And, in general, given a sample of appreciable size, the mode of a posterior distribution based on an informationless prior will never differ greatly from the ML estimate of the parameter in question, and they often coincide (as in Example 7 below).

EXAMPLE 6 (Benford's law). In tables of numerical data, such as one finds in an Atlas or Census Report, what is the probability p_k that k occurs as first significant digit, $k = 1, \ldots, 9$? An application of the classical principle of indifference suggests that all nine digits should occur about equally often as initial digit. The physicist, Frank Benford, found, however, that they do not, and this has been taken as further evidence that probability distributions can be securely based only on frequency data.[17] Suppose, nevertheless, that there is a definite distribution governing the frequencies in question. Let X result from a tabular entry by writing a decimal point immediately following the first significant digit. Then X lies in the interval $[1, 10]$, and the first significant digit is k iff X lies in the subinterval $[k, k+1]$. Since nothing is said about the scale in which the tabular entries are expressed, the distribution of X should be scale invariant, hence log-uniform.[18] In particular, $p_k = [\log(k+1) - \log k]/\log 10$, and by taking logarithms to the base 10 we arrive at:

$$(3.13) \qquad p_k = \log(k+1) - \log k, \, k = 1, \ldots, 9.$$

The distribution thus obtained is precisely the one Benford conjectured on empirical grounds upon examining numerous tables! It is far more insightful, however, to derive the distribution from considerations of invariance. That derivation tells us, for example, that any table which failed to obey the distribution would be such that a change of scale would appreciably alter the frequencies with which $1, \ldots, 9$ occur as initial digit. It is illuminating, in this connection, to start with an equally spaced sequence, like 5, 10, 15, 20, $\ldots, 95$, which manifestly violates Benford's law, and find, upon multiplying the entries successively by different factors, that the resulting sequences fit Benford's distribution better and better. Other less convincing explanations of the law, including Benford's, are discussed by Raimi (1969), who credits the observation that scale invariance uniquely determines the distribution (3.13) to R. S. Pinkham. As an example of its empirical goodness-of-fit, I looked up a table in my Atlas giving the areas of the world's 124 largest islands. The observed and expected relative frequencies follow:

TABLE VI

	1	2	3	4	5	6	7	8	9
Obsv	0.298	0.218	0.153	0.089	0.064	0.016	0.073	0.056	0.032
Exp	0.301	0.176	0.125	0.097	0.079	0.067	0.058	0.051	0.046

As you can see, the fit is excellent. We are loath to believe that agreement this good — and this improbable — could be mere coincidence.

EXAMPLE 7. (Bernoulli trials). The binomial law (3.1) gives the probability of r successes in n Bernoulli trials. We seek the informationless prior for p, the success rate. To this end, Jaynes (1968) develops an especially interesting invariance argument. Since $f(p)dp$ is the probability of a probability, we imagine that $f(p)$ represents, not the partial knowledge of any one person, but the distribution of beliefs held by a large population of diverse individuals. It is possible, as Jaynes notes, that each individual hold a definite opinion but that the population as a whole be in a state of 'total confusion' regarding p. Now he imagines that an additional item E of information is given to each man before the first trial. Using Bayes' rule, Mr X, who previously believed $p = P(S/X)$ to be the probability of success, will change his estimate to

$$p' = P(S/E, X) = \frac{P(S/X)P(E/SX)}{P(E/SX)P(S/X) + P(E/FX)P(F/X)}$$

where $P(F/X) = 1 - P(S/X)$ is his initial estimate of the probability of failure. Thus, E induces a mapping of the parameter space, $0 \leqslant p \leqslant 1$, onto itself, given by:

(1) $\qquad p' = ap/(1 - p + ap)$, where $a = P(E/SX)/P(E/FX)$.

To say the population is in a state of total confusion regarding p is to say it can learn nothing from the additional information E. That is, after receipt of E, the total number of individuals who believe p is in any given interval (p_1, p_2) is the same as before.

Mathematically, the original distribution of beliefs $f(p)$ is shifted by the transformation (1) to a new distribution $g(p')$ with

(2) $\qquad f(p)\mathrm{d}p = g(p')\mathrm{d}p'$

and the invariance condition (that the population learned nothing) requires that f and g be identical functions:

(3) $\qquad f(p) = g(p)$.

Combining (1)–(3),

$$f(p)\mathrm{d}p = g\left(\frac{ap}{1 - p + ap}\right)\frac{a\,\mathrm{d}p}{(1 - p + ap)^2}$$

$$= af\left(\frac{ap}{1 - p + ap}\right)\frac{\mathrm{d}p}{(1 - p + ap)^2}$$

or

(4) $\qquad (1 - p + ap)^2 f(p) = af\left(\frac{ap}{1 - p + ap}\right)$.

Differentiating both sides of (4) with respect to a gives:

$$D_a(1 - p + ap)^2 f(p) = 2p(1 - p + ap)f(p),$$

$$D_a \left[af\left(\frac{ap}{1 - p + ap} \right) \right] = f\left(\frac{ap}{1 - p + ap} \right)$$

$$+ af'\left(\frac{ap}{1 - p + ap} \right) D_a\left(\frac{ap}{1 - p + ap} \right),$$

$$= f\left(\frac{ap}{1 - p + ap} \right)$$

$$+ af'\left(\frac{ap}{1 - p + ap} \right)\left(\frac{p(1 - p)}{(1 - p + ap)^2} \right),$$

whereupon, setting $a = 1$ gives the differential equation:

(5) $(2p - 1)f(p) = p(1 - p)f'(p),$

with general solution:

(3.14) $f(p) = cnst \times p^{-1}(1 - p)^{-1},$

the (improper) Beta density $cnst \times p^a(1 - p)^b$ with both parameters, a and b, equal to -1. This becomes a proper density upon observing one success and one failure. Against this improper prior, the Bayes estimate of p based on a sample of r successes in n trials is just r/n, the observed success rate (and ML estimate of p). This result confirms the informationless character of the prior (3.14).

Given a beta prior

(3.15) $\text{Beta}(a, b) = \dfrac{a!b!}{(a + b + 1)!} p^a(1 - p)^b$

with parameters a and b, and given s successes and f failures in a sequence of $n = s + f$ Bernoulli trials, the posterior distribution of p is again beta, in fact, Beta $(a + s, b + f)$. This additivity of the prior and sample parameters gives particularly vivid expression to the averaging of prior beliefs and sample information performed by Bayes' rule. The mean of Beta (a, b) is

(3.16) $(a + 1)/(a + b + 2),$

and the variance is

(3.17) $$\frac{(a+b)(b+1)}{(a+b+2)^2(a+b+3)}.$$

If both parameters are equal, say to $a-1$, the beta distribution becomes symmetrical about its mode (or mean) $1/2$. Also, since a binomial random variable can be regarded as taking the values 1 or 0 with probabilities p and $1-p$, and $1 \cdot p + 0 \cdot (1-p) = p$, we see that the probability of success on the next trial is equal to the mean of the posterior density of p. If the prior is Beta $(a-1, a-1)$, the posterior mean in question, based again on s successes and f failures, is the mean of Beta $(a-1+s, a-1+f)$:

(3.18) $\qquad (a+s)/(2a+n), n = s+f.$

But (3.18) is the general expression given in Carnap (1952) for the predictive probability of success on the next trial of a Bernoulli sequence that begins with s successes and f failures. It is tantamount to using a weighted average of the a priori estimate of p, $1/2$, and the sample success rate, s/n, to estimate the probability of success on the next trial, where $2a{:}n$ is the ratio of the weights accorded these two factors (which Carnap calls, resp., the 'logical factor' and the 'empirical factor'). The parameter a may therefore be considered a *learning rate parameter* or a *flattening constant* (Good, 1965). Carnap's parameter λ is related to my a by $\lambda = 2a$.

There is an obvious relation between Carnap's 'continuum of inductive methods', given by (3.18), and the family of beta priors. We have already seen that a Bayesian who starts with the symmetric prior, Beta $(a-1, a-1)$, will arrive at the same predictive probabilities, given a Bernoulli sequence, as a Carnapian who chooses the flattening constant $a = \lambda/2$ in (3.18). The converse follows from De Finetti's celebrated Representation Theorem. Loosely expressed, this theorem asserts that a subjectivist for whom the probability of s successes and f failures in a sequence of dichotomous trials depends only on s and f and not on the trial numbers on which successes occur (i.e., for whom the trials are 'exchangeable') will arrive at predictive subjective probability assessments indistinguishable from the predictive probabilities of an objectivist who views the trials as Bernoullian and has a (uniquely determined) prior probability distribution over p, the constant probability of success. Hence, one who arrives at the predictive probability (3.18) on the basis of n exchangeable trials would have arrived there had he viewed the trials as Bernoullian (independent with stationary probability of success) and adopted Beta $(a-1, a-1)$ as his prior distribution of p. Since the uniqueness part of De Finetti's theorem tells us that only one prior will lead our objectivist to

(3.18), and Beta $(a - 1, a - 1)$ does lead there, Beta $(a - 1, a - 1)$ must be that prior. In fine, use of the flattening constant a in (3.18) corresponds biuniquely to adoption of the symmetric beta prior, Beta $(a - 1, a - 1)$. By running the experiment backwards, so to speak, the prior density of p can be obtained from the sample outcome and the posterior mean — provided, of course, that the trials are exchangeable.

Carnap does assume exchangeability, which, in his terminology, translates into the assumption that all the state descriptions which belong to a given structure description have equal probability. Yet, he apparently failed to see the connection between his 'continuum' and the family of symmetric beta priors (at least he never mentions it), and this failure led him to say some very strange things. Much of his discussion (esp. Section 18 or Carnap (1952)) makes it sound as though one chooses a particular value of λ for life; each such choice is apparently thought of as tantamount to a choice of 'inductive method'. In the same vein, Carnap proffers broad general criticisms of different such 'methods' taken *in abstracto* (*op. cit.* Sections 13, 14). E.g., he criticizes the 'extreme' methods, $\lambda = 0$ (which corresponds to the improper beta prior, Beta $(-1, -1)$, to which Jaynes' invariance argument led us) and $\lambda = \infty$ (which corresponds to the beta prior which concentrates the entire probability mass at $p = 1/2$). The latter is criticized for being 'impervious to experience'; the estimate of p remains fixed at $1/2$ no matter what the sample outcome.

Now this criticism is seen to be misguided once the correspondence between values of λ and beta priors is recognized. The degenerate beta prior, Beta (∞, ∞), which concentrates all the probability mass at $p = 1/2$, is tantamount to having observed an infinite number of trials with equal numbers of successes and failures. Such a prior is appropriate only for one who knows $p = 1/2$. Given such knowledge, an inductive policy impervious to observed frequency counts seems entirely reasonable.

The beta priors, one improper and one a delta function, which correspond to the 'extreme' methods, $\lambda = 0$ and $\lambda = \infty$, are not mere 'formal elements' adjoined to remove singularities of the biunique correspondence. They represent complete ignorance and complete knowledge, respectively, and these facts must control the interpretation of calculations made with the associated Carnapian 'methods' — not vice versa. E.g., one can 'formally' compute the predictive probability of success based on the informationless prior, Beta $(-1, -1)$, and a sample devoid of failures, by plugging $a = 0$ into the expression (3.18). The probability is then equal to one, of course, and consequently, the probability that failures will be absent from any future sequence, finite or infinite, remains one. Strictly speaking, however, the density of p remains improper until both a success and a failure have been observed. I interpret this to mean that if one starts out in a state of complete ignorance regarding p, then, while observation of 'mixed' sequences containing both successes and failures will lead to determinate beliefs about p, no determinate beliefs will be forthcoming if failures (resp. successes) are absent. Lacking prior theoretical grounds for singling out one of the extreme values 0, 1, no finite 'unmixed' sequence of trials can lead one to assign a determinate probability to one of these values.

On the other hand, if there are grounds for singling out $p = 1$, say, entirely different methods of analysis become appropriate. One should then test $p = 1$ against the composite alternative $p < 1$ by the method of Chapter 5, viz., by computing the ratio of average likelihoods, which in this case is easily seen to be:

$$L = \left(\int_0^1 p^s \, dp \right)^{-1} = s + 1 .$$

That is, observation of s successes and 0 failures would multiply the prior odds in favor of $p = 1$ by a factor of $s + 1$. To take an especially clear case, the experiment might consist in flipping a coin drawn from a population of coins known to contain a given proportion of two-headed coins, as well as various other bent coins, our knowledge of the bent coins justifying a flat distribution over the values of $p \neq 1$. This state of knowledge is very far removed from complete ignorance.

In any event, there is no 'uniformly best' inductive method in the sense of Carnap, nor even the possibility of ruling some of them out on a priori grounds. All that can be said is that different values of Carnap's λ correspond to different states of prior knowledge, and consequently, different 'inductive methods' (i.e., beta priors) are appropriate in different contexts.

EXAMPLE 9 (Bertrand's Paradox). If Bertrand's problem can be settled by an invariance argument, a not inconsiderable part of the traditional case against Bayesian methods collapses. Jaynes (1970) has supplied the argument, and I sketch it herewith.

The problem is to find the probability that a chord of a circle 'drawn at random' exceeds a side of the inscribed equilateral triangle. As seen from Figure 1, a side of this inscribed triangle has length $\sqrt{3}R$, where R is the radius. The difficulty is that by using different geometric constructions for the chord, and imposing uniform priors on different parameters, different

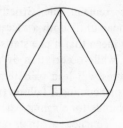

Fig. 1.

answers to Bertrand's question are forthcoming. Thus, if (r, θ) is the midpoint of the chord in polar coordinates, one could assign a uniform prior to this point itself, or to its distance from the center, or to the angle which the chord forms with the circumference, etc. Each of these methods leads to a different result, and there seems to be little or nothing to choose between them. However, the appearance that the problem is ill-posed is deceptive in this case, and, again, 'indifference between problems' provides the missing clue. For nothing is said about the size, position, or orientation of the circle in the statement of the problem.

To help fix our ideas, imagine a physical embodiment of the problem: broom straws are tossed onto a circle drawn on the floor. Now the circle must be sufficiently small to preclude *skill*, and so the implied requirement of *scale invariance* is that *small* changes in the size of the circle (or its radius) leave the problem unchanged. Similarly, small displacements of the circle's center should leave it unchanged. And, finally, the orientation of the circle (or the direction from which the straws are tossed) shouldn't matter. Each of the symmetries implied in the statement of the problem – rotational, scale, and translational – imposes a constraint on the prior density $f(r, \theta)$ of the midpoint (r, θ) of the chord. First, it is obvious by rotational symmetry that $f(r, \theta) = qr^{q-2}/(2\pi R^q)$, $q > 0$. Finally, by translational invariance, $q = 1$, whence the density is:

$$(3.19) \qquad f(r, \theta) = \frac{1}{2\pi rR} \ .$$

From Figure 1 it is also seen that the chord exceeds a side of the inscribed equilateral triangle iff its midpoint is a distance $r < R/2$ from the center. Hence the answer to Bertrand's question is given by

$$\int_0^{2\pi} d\theta \int_0^{R/2} \frac{1}{2\pi rR}\, r\, dr = 1/2 \ .$$

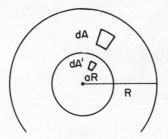

Fig. 2.

As Jaynes points out, translational invariance alone uniquely determines the density (3.19), but it is necessary in any case, to check the consistency of this requirement with that of scale invariance. (Rotational invariance is obviously consistent with either of the others.) Had the solutions yielded by scale and translational invariance not been in agreement, the problem would lack a solution for being *overdetermined*.

Jaynes reports having carried out the actual physical experiment with straws and finding the results in good agreement with (3.19). Poor agreement between theory and data would imply (as in the other examples) that one of the implied symmetries did not hold.

I now give the details of Jaynes' argument. I assume $f(r, \theta) = f(r)$ is independent of θ (the proof from rotational invariance being trivial).

Scale Invariance. Two concentric circles of radii R and aR, $0 < a < 1$, are shown in Figure 2. Two corresponding elements of area, $dA = r\, dr\, d\theta$ and $dA' = ar\, d(ar)d\theta$ are also shown. Let $f(r)r\, dr$ be the probability that (r, θ) lies in dA. If h is the corresponding density for the smaller circle, we have the transformation condition

$$(1) \qquad f(r) = h(r)2\pi \int_0^{aR} f(r)r\, dr, \quad 0 \leqslant r \leqslant aR,$$

by applying the definition of conditional probability for densities. According to (1), the probability that (r, θ) lies in $dA' = ar\, d(ar)d\theta$ is the probability $2\pi \int_0^{aR} f(r)r\, dr$ that it lies in the smaller circle multiplied by the conditional probability $h(r)$ that it lies in dA' given that it lies in the smaller circle. By scale invariance, two corresponding elements of area, dA and dA', are related to the large and small circles respectively in the same way, and so they must be assigned equal probabilities by the respective densities f and h:

$f(r)r\ dr = h(ar)ar\ d(ar) = a^2h(ar)r\ dr$, whence we obtain the symmetry condition:

(2) $f(r) = a^2 h(ar)$.

Combining (1) and (2), $a^2 h(ar) = 2\pi h(r)\int_0^{aR} f(u)u\ du$. But, by (1), $f(r) \propto h(r)$ in the smaller circle, whence

(3) $a^2 f(ar) = 2\pi f(r) \int_0^{aR} f(u)u\ du$.

The most general solution f of the functional Equation (3) which integrates to one is

(4) $f(r) = \dfrac{qr^{q-2}}{2\pi R^q}$, $q > 0$.

as the reader may easily verify.

   Translational Invariance. Let C and C' be two circles with centers a small distance d apart (Figure 3). If a straw falls across both circles, its midpoint lies at a point $P(r, \theta)$ with respect to circle C and at a corresponding point $P'(r', \theta')$ with respect to circle C'. The coordinates of P and P' are related by the transformation

(5) $r' = |r - b\cos\theta|$ and $\theta' = \begin{cases} \theta \text{ if } r > b\cos\theta , \\ \theta + \pi, \text{ if } r < b\cos\theta , \end{cases}$

with Jacobian equal to one. Now translation invariance implies that the probability with which the midpoint of the chord falls respectively in

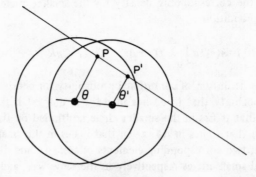

Fig. 3.

corresponding neighborhoods Λ and Λ' of P and P' should be the same. These probabilities are:

$$(6) \qquad f(r) r \, dr \, d\theta = \frac{q}{2\pi R^q} \int_\Lambda r^{q-1} \, dr \, d\theta$$

and

$$(7) \qquad \frac{q}{2\pi R^q} \int_{\Lambda'} (r')^{q-1} \, dr' \, d\theta' = \frac{q}{2\pi R^q} \int_{\Lambda'} |r = b \cos \theta \,|^{q-1} \, dr \, d\theta$$

using the fact that the Jacobian of the transformation (5) is one. But (6) and (7) agree iff $q = 1$.

7. CONCLUSION

Expressing what was no doubt a widespread consensus at the time, Good (1968) wrote that Bayesian methods "achieve a greater reduction of the data at the sacrifice of some objectivity". I am not sure whether many Bayesians today, including Good, would be at ease with this way of putting it. Certainly Bayesian methods achieve greater reduction and have wider application than the methods of sampling theory.[19] But it has become less clear that objectivity is thereby lost. Nevertheless, the impression that Bayesian methods are indelibly subjective persists. The objectivist Bayesian position I have sketched in this chapter may be viewed as an attempt to redraw the lines of the Bayesian debate. Criticisms of subjectivist formulations simply do not touch the objectivist Bayesian position. To merit serious consideration, however, this position must present an adequate treatment of informationless priors.

Objections to a probabilistic representation of ignorance have long been thought devastating. These objections are forcefully re-stated in Fine (1973). Let me briefly summarize the main points of Fine's discussion, especially as they bear on the work of Jaynes and Jeffreys.

(i) Maximum entropy distributions do not always exist.

(ii) Invariance arguments often lead to improper priors.

(iii) It is naive to suppose that every state of knowledge determines a unique probability distribution.

(iv) Probability distributions can be securely based only on frequency data.

(v) Parametrization is arbitrary, and so any attempt to assign prior distributions on the basis of ignorance is bound to generate inconsistency.

The reader who has followed the discussion of the preceding sections will see that we have gone some way towards meeting each and every one of these criticisms.

First, where maximum entropy distributions do not exist, natural extensions of the method may suggest themselves, as I illustrated at the end of Section 5. Regarding (ii), improper priors are to be regarded as local approximations (over the region where the likelihood function is substantial) to a proper prior (which is flat in the region of interest, but tails off sufficiently rapidly outside this region to integrate). Any proper prior with these properties will lead to essentially the same posterior distribution ('stable estimation'). It *is* naive to suppose that every state of knowledge determines a probability distribution, but it is equally naive to think that probability distribtions can be based only on frequency counts. Fine cites Benford's law in this connection because, naively, one expects all nine digits $1, \ldots, 9$ to occur about equally often as first digit in a table of numerical data. Since this does not occur, it is taken as a strike against the classical principle of indifference. Yet, as we saw in Example 6 above, scale invariance determines the correct distribution ('Benford's law'). It is less than clear, consequently, that probability distributions like Benford's law, the distributions of statistical mechanics, or the Jaynes distribution for the Bertrand problem (Example 8), can be securely based only on frequency data. Even those who are sympathetic to empirical theories of probability must, it seems to me, take seriously the claims of a method which, to put it bluntly, *works*. Moreover, there is a clear and compelling explanation for why it works, as described in Section 5.

Faced with the Bertrand problem, Jaynes asks: "Do we really believe it is beyond our power to predict by *pure thought* the result of such a simple experiment?" Someone confronted with a Galton board might ask the same question. To be sure, the appropriate probability distribution is not in either case produced by thought alone. Empirical assumptions are made, or, if you like, it is only an idealized problem for which a probability distribution is adduced. No matter, for all of science proceeds in like fashion. And, as these examples suggest, the distinction between 'direct' and 'prior' probabilities is largely spurious anyway. 'Direct' probabilities are no more inherently public or objectively grounded than prior probabilities. Considerations of scale-invariance led us to the log-uniform distribution for the frequencies of first digits; the very same considerations underlie the log-uniform prior for a scale parameter (Examples 4,5). The fact that we are led in the one case to a data distribution and in the other to a prior distribution seems of scant relevance. Indeed, as in the birthday problem, whether a probability counts as 'direct' or 'prior' is largely determined by the use to be made of it. The issue is less whether prior probabilities are objective or subjective than whether the theory of probability is to be limited to 'empirical probabilities' (however specified). The situation here is not unlike that in the foundations of mathematics where intuitionists would impose limitations on the basis of a critique of classical analysis that seems anything but compelling.

I want to conclude with some additional remarks on parametrization and ignorance. One could take the position that all variables are discrete (that measurements are of limited precisions, etc.). In that case, the paradoxes associated with parameter transformation would all disappear. I think such a position quite defensible. Still it would be nice to be able to show that a consistent treatment of continuous variates can be given, and, in particular, an adequate treatment of informationless priors for continuous parameters.

One may, with E. T. Jaynes, view states of complete ignorance as entering probability theory as naturally as the number zero enters arithmetic. On the other hand, one may, with Box and Tiao (1973), think of ignorance, more pragmatically, as the state of having little information relative to what the contemplated experiment is expected to deliver. In the latter conception, informationless priors for the same parameter will be expected to vary with the experiment or sampling scheme. Moreover, if one is ignorant in this relative sense, a change of parameter will be anything but 'arbitrary'. One who is satisfied with a prior that assigns equal weight to θ lying in the intervals $[4, 5]$ and $[5, 6]$ will not be indifferent to the reparametrization of the problem which replaces θ by θ^{10}, for then almost six times as much weight would be accorded the latter interval.

But, as Box and Tiao also point out (pp. 26–27), reparametrization need not be a matter of indifference even if one is ignorant in a stronger sense. They show, for example, that if one is ignorant of the mean θ of a normal population, with known variance, it makes all the difference whether one assigns a flat prior to θ itself or to θ^{-1}. Our state of knowledge is characterized in this case by saying that the data should provide information about the location of the population, but not about the already known spread. In terms of θ^{-1}, however, the likelihood curves corresponding to different observed sample means will have different spreads, as well as different locations.

One can approach the problem of ignorance from the standpoint of transformation groups, but there is another approach, also due originally to Jeffreys (1961). Given an informationless or 'neutral' prior, the posterior distribution of θ should depend only on the likelihood function.

In that case, the posterior distribution is obtained simply by normalizing the likelihood function (so that it integrates to one.) Now the likelihood function, under weak regularity conditions, has an approximately normal shape, and, as we will see in Chapter 5, the normalizing constant, multiplication by which turns the likelihood into a true probability density, is proportional to the square root of the determinant I of Fisher's information matrix (Chapter 1). A prior proportional to $I^{1/2}$ will thus be informationless in the required sense. Moreover, I has (Chapter 1) the fundamental property:

$$(3.20) \qquad I(\phi) = I(\theta) \left(\frac{d\theta}{d\phi} \right)^2, \qquad \text{where } \phi = \phi(\theta) \text{ is a } 1-1 \text{ transformation of } \theta.$$

Hence, $p(\theta)d\theta \propto I^{1/2}(\theta)d\theta \propto I^{1/2}(\phi)d\phi \propto p(\phi)d\phi$, and so the Jeffreys prior (proportional to $I^{1/2}$) is invariant under $1-1$ transformations of the parameter.

Being invariant, one would expect priors given by Jeffreys' rule to agree with those obtained earlier by invariance arguments involving 'indifference between problems'. Happily, the two methods almost always agree. The informationless prior for the parameter p of a binomial distribution is the single exception known to me. In this example, the Jeffreys prior is $\propto p^{-1/2}(1-p)^{-1/2}$, while the Jaynes prior (Example 7) is $\propto p^{-1}(1-p)^{-1}$, the latter being improper. Here, though, the Jaynes prior was not actually based on 'indifference between problems', but on an ingenious but somewhat artificial construal of total ignorance in terms of 'total confusion'. One feels justified, then, in saying that invariance arguments based squarely on 'indifference between problems' are in substantial agreement with the method of Jeffreys.

Confluence of different explications is always strongly indicative of their soundness. The equivalence of variously motivated explications of 'effectively computable' is a striking case in point. Here we have *substantial* confluence of two explications of ignorance. As matters turn out, there is a third representation of ignorance, due to Box and Tiao (*op. cit.*) which may well be thought to supersede the other two, while concurring with them in a broad spectrum of cases.

Their method begins by assuming a uniform (or 'locally uniform') prior for a location parameter. For such parameters, the likelihood function is *data translated*, that is, determined up to location by the sample datum. Given any parameter θ, one then seeks a $1-1$ transformation $\phi = \phi(\theta)$ such that the likelihood is data translated when expressed in ϕ. Then ϕ should be locally uniform, and this distribution of ϕ determines a (typically non-uniform) distribution of θ, in fact, a distribution proportional to $|d\phi/d\theta|$. For example, if σ is the standard deviation of a normal population, $\log \sigma$ is easily seen to be data translated, and so it should be locally uniform. Hence, taking $\phi = \log \sigma$, ϕ is log-uniform. The treatment of many parameters is similar, and I omit the details.

For one-parameter distributions, the method of Box and Tiao agrees with the Jeffreys rule, but the methods can diverge in multiparameter problems. As Jeffreys himself pointed out, his method yields a joint prior $\propto \sigma^{-2}$ for the mean and standard deviation of a normal family, while, if the two parameters are known to be independent, separate consideration leads to the joint density $\propto \sigma^{-1}$, the prior to which the method of Box and Tiao leads. The Jeffreys rule is clearly superseded in this case, for it ignores the assumed independence of the two parameters. The point is discussed in Box and Tiao (1973), Section 1.3.6.

Critics of the Bayesian position have been right to focus on the problem of uninformative priors.[20] It should be clear from the discussion, however, that rather sophisticated representations of ignorance are being evolved. The methods of Jaynes, Jeffreys, and Box and Tiao which I have sketched seem fully capable of handling the difficulties traditionally raised. The sincerity of ancient and recent attempts to demolish the Bayesian position can scarcely be doubted, and one can only agree with Sir Karl Popper's contention (cf. Chapter 6) that a theory 'proves its mettle' when repeated attempts to overthrow it fail. At the present juncture the criticisms that have been lodged seem insufficient to justify limiting probability theory to problems for which frequency data is available. Even less is gained by a retreat to subjectivism which, in practice, hardly differs from mere reporting of the observed likelihood function. Nevertheless, the reader must judge for himself whether the (objectivist) Bayesian 'research programme' outlined in this chapter is 'progressive' or 'degenerating'.

NOTES

[1] Peirce (*Collected Papers*, 2.649) describes as 'predictively reliable' those inference rules which lead a high proportion of the time from true premises to true conclusions, citing Locke as his source.

[2] A 'reasonable argument' in the sense of Adams (1966) is one which never leads from highly probable premises to highly improbable conclusions. But such arguments are not ampliative and their semantics (as developed by Adams) is probabilistic. Therefore, they afford little aid and comfort to inductivism.

[3] Feigl (1950) and Reichenbach (1938), Sections 38–39. The pragmatic 'vindication' and the Strawsonian 'dissolution' are both anticipated in Ramsey (1926).

[4] Strawson (1952), pp. 256–257.
[5] The concept of instancehood alluded to is, of course, that of Hempel (cf. Hempel (1965), Chapter 1).
[6] For mathematical details, cf. Feller (1957), pp. 109ff.
[7] A justification in terms of rational betting is given in Kemeny (1955); another, close to the spirit of our own, in Teller (1976).
[8] Savage (1954), Chapter 4, esp. p. 57 and pp. 63–68.
[9] Jaynes himself has applied the principle to the derivation of the laws of statistical mechanics; cf. Note 13 below.
[10] The 'widget' problem discussed in Jaynes (1963) is a good illustration.
[11] As in Feller (1957), p. 39: "We have here an instructive example of the impossibility of justifying probability models by a priori arguments ... no pure reasoning could tell that photons and protons would not obey the same probability laws".
[12] Compare Williams (1947).
[13] The original papers are Jaynes (1957). A good recent exposition is given in Jaynes (1967).
[14] As pointed out by Fine (1973), p. 173.
[15] Jaynes (1968), pp. 235ff.
[16] Cf. Edwards et al. (1963).
[17] Fine (1973), p. 168.
[18] Compare the derivation in Feller (1966), pp. 62–63.
[19] See Chapter 10.
[20] As in Edwards (1972), Chapter 4.

BIBLIOGRAPHY

Adams, E.: 1966, 'Probability and the Logic of Conditionals', *Aspects of Inductive Logic* (K. J. .J. Hintikka and P. Suppes, eds.), North-Holland Publ. Co., Amsterdam.

Box, G. E. P. and Tiao, G. E.: 1973, *Bayesian Inference in Statistical Analysis*, Addison-Wesley, Reading, Mass.

Carnap, R.: 1952, *The Continuum of Inductive Methods*, University of Chicago Press, Chicago.

DeFinetti, B.: 1969, 'Initial Probabilities: A Pre-requisite to Any Valid Induction', *Synthese* 20, 1–13.

DeFinetti, B.: 1972, *Probability, Induction and Statistics*, Wiley, N.Y., Chapters 1–3, 8, 9.

Edwards, A. W. F.: 1972, *Likelihood*, Cambridge University Press, Cambridge.

Edwards, W., Lindman, H., and Savage, L. J.; 1963, 'Bayesian Statistical Inference for Psychological Research', *Psych. Rev.* 70, 193.

Feigl, H.: 1950, 'De Principiis Non-disputandum . . . ?', *Philosophical Analysis* (M.Black, ed.), Prentice-Hall, Englewood Cliffs, New Jersey.

Feller, W.: 1957, *Introduction to Probability Theory and its Applications*, I, 2nd ed., Wiley, New York.

Feller, W.: 1966, *Introduction to Probability Theory and its Applications*, II, Wiley, New York.

Fine, T.: 1973, *Theories of Probability*, Academic Press, New York.

Gibbs, J. W.: 1902, *Elementary Principles in Statistical Mechanics*, Yale University Press, New Haven.

Good, I. J.: 1965, *The Estimation of Probabilities*, MIT Press, Cambridge, Mass.

Harper, W. A.: 1976, 'Rational Belief Change', *Foundations of Probability Theory*,

Statistical Inference, and Statistical Theories of Science (W. A. Harper and C. K. Hooker, eds.), II, D. Reidel, Dordrecht.

Hartigan, J. A.: 1964, 'Invariant Prior Distributions', *Ann. Math. Stat.* 35, 836.

Hume, D.: 1748, *An Enquiry Concerning Human Understanding* London.

Jaynes, E. T.: 1957, 'Information Theory and Statistical Mechanics', Part I, *Phys. Rev.* 106, 620–630; Part II, *ibid.*, 108, 171–191.

Jaynes, E. T.: 1963, 'New Engineering Applications of Information Theory', *Proc. 1st Symp. on Engineering Appls. of Random Function Theory and Probability* (J. L. Bogandoff and F. Kozin, eds.), Wiley, N.Y., pp. 163–203.

Jaynes, E. T.: 1967, 'Foundations of Probability Theory and Statistical Mechanics', *Delaware Sem. in the Foundations of Physics* (M. Bunge, ed.), Springer, Berlin.

Jaynes, E. T.: 1968, 'Prior Probabilities', *IEEE Trans. Systems Sci. Cybernetics*, SSC–4, pp. 227–241.

Jaynes, E. T.: 1970, 'The Well-Posed Problem', *Foundations of Statistical Inference* (V. P. Godambe and D. A. Sprott, eds.), Holt, Rinehart and Winston, Toronto.

Jeffrey, R. C.: 1965, *The Logic of Decision*, McGraw-Hill, New York.

Jeffreys, H.: 1961, *Theory of Probability*, 3rd ed., Oxford University Press, Oxford.

Keynes, J. M.: 1921, *A Treatise of Probability*, Macmillan, London (reprinted Harper Torchbooks, N.Y., 1962).

Peirce, C. S.: 1878, 'The Probability of Induction', *Collected Papers*, II (C. Hartshorne and P. Weiss, eds.), Harvard Press, Cambridge, Mass., pp. 415–432.

Peirce, C. S.: 1883, 'A Theory of Probable Inference', *ibid.*, 433–477.

Raimi, R.: 1969, 'The Peculiar Distribution of First Digits', *Sci. Am.* 221, no. 6, 109–120.

Ramsey, F. P.: 1926, 'Truth and Probability', reprinted in *Studies in Subjective Probability* (H. Kyburg and H. Smokler, eds.), Wiley, New York.

Reichenbach, H.: 1938, *Experience and Prediction*, Chicago.

Savage, L. J.: 1954, *The Foundations of Statistics*, Wiley, New York.

Savage, L. J.: 1971, 'Elicitation of Personal Probabilities and Expectations', *J. Am. Stat. Assoc.* 66, 783–801.

Schmitt, R.: 1969, *Measuring Uncertainty*, Addison-Wesley, Reading, Mass.

Strawson, P. F.: 1952, *Introduction to Logical Theory*, Wiley, New York.

Teller, P.: 1976, 'Conditionalization, Observation and Change of Preference', *Foundations of Probability Theory, Statistical Inference, and Statistical Theories of Science*, II, D. Reidel, Dordrecht.

Williams, D. C.: 1947, *The Ground of Induction*, Harvard Press, Cambridge, Mass.

CHAPTER 4

INDUCTIVE GENERALIZATION

Given the number of kinds present in a sample, and their frequencies, we may wish to infer the number of kinds present in a population. A paradigm is to estimate the number of words (types, not tokens) in a book given a sample of text.

By a *composition* of a population I mean an inventory of the kinds present, together with their frequencies. A strict Bayesian solution of our problem would entail writing down all possible compositions, assigning them prior probabilities, then using the multinomial distribution to calculate the probability of the observed sample conditional on each composition. Even for moderately large populations, the amount of calculation involved is inordinate. Besides, we cannot always list in advance all possible kinds we might encounter. An approximate solution is clearly necessary.

Hintikka has proposed replacing compositions by *constituents*. These specify which of k possible kinds are present in the population, but not their frequencies. For its prior probability, he assigns a constituent its chance of holding in a finite random universe of size α. (A 'random universe' is one in which individuals are assorted among kinds at random.) There are several ways in which the likelihoods might be specified, but the details needn't concern us, for almost any reasonable choices would lead to the same qualitative result Hintikka obtains: that the simplest constituent consistent with the sample (viz. that which designates all and only those kinds present in the sample) becomes overwhelmingly probable at large sample sizes, the rate of convergence to certainty being controlled by the prior distribution. This is but one illustration of a general phenomenon: the tendency of the likelihood function to concentrate in small neighborhoods of its maximum at large samples.

The considerable insight it affords notwithstanding, Hintikka's approximate solution is scarcely more applicable to real problems than the strict Bayesian approach. It has, in addition, a purely conceptual limitation which Professor Hintikka himself emphasized in a recent seminar at Stanford (March, 1975). The learning rate parameter α is meant to reflect one's beliefs about the entropy of the population (entropy of a composition increasing with the number of kinds and the uniformity of their frequency distribution).

As such, α should be subject to revision in the light of sample data. Having fixed the value of α, however, you are stuck with it. While deploring this feature of his system, Hintikka went on to express doubt whether a learning rate parameter could itself be revised by conditionalizing on the sample. I put forward a proposal to that effect at the seminar, which has since evolved into the present formulation.

The sampling of a population for species is itself a kind of stochastic process, and one that is most conveniently studied by considering the *rate* at which new kinds are appearing in successive samples. Let r_n be the proportion of new (i.e. previously unsampled) kinds in the nth sample of a sequence, each sample being of the same size. What can we say about the sequence of r_n?

EXAMPLE 1. There are 100 kinds K_i of which the first ten 'common' kinds occur with frequency $9/100$, while the remaining 90 'rare' kinds all occur with frequency $1/900$. If we take successive samples of 20, the common kinds will be expected about twice (viz. 1.8 times), while the rare kinds are expected once in 45 samples. As a first approximation, assume that 90% of the items in any sample are common and 10% rare, reflecting the population frequencies. Then the probability that m of the 10 common kinds are absent in a sample is

$$p_m(18, 10) = \binom{10}{m} \sum_{i=0}^{10-m} (-1)^i \binom{10-m}{i} (1 - (m+i)/10)^{18} .$$

E.g., $p_0(18, 10) = 0.134673$, $p_1(18, 10) = 0.385289$, $p_2(18, 10) = 0.342987$, $p_3(18, 10) = 0.119425$, etc., whence the probability that more than three of the common kinds are absent is 0.017626. The most probable value of r_1 is $11/20$, with one common kind absent. The missing common kind is likely to be present in the second sample (with probability $1 - 0.9^{18}$). And by our assumption, 18 of the 20 items in the second sample will be common kinds, while, of the 2 items of rare kinds, it is unlikely that either of those rare kinds will have appeared in the first sample. The most probable value of r_2 is therefore $3/20$. At this point, all common kinds have appeared, hence the only new kinds in future samples are rare kinds. Since there are many of these, the subsequent average decrease of the r_n will be much slower. The reader can work out other cases for himself. In every case, on the average, the r_n should decrease at a diminishing rate which is a function of the entropy of the true composition of the population, the higher the entropy, the slower the rate of decrease.

The first part of the problem is to specify a rich and tractable family of convex decreasing functions. I tentatively propose the family $r_n = Me^{-n^t}, M, t > 0$. The second part is to single out a member of this family which, in some sense, fits the observed r_n best. I will consider two methods. Having singled out a member of the family as best-fitting, the third part of the problem is to base an estimate of the number of kinds on it. This occasions no great difficulty, and I will have little to say about it.

EXAMPLE 2. I took an English translation of Per Lagerqvist's novel *The Dwarf* and divided the first 1000 words into ten blocks of 100 words each. The number of new words per block was 75, 52, 30, 39, 40, 28, 26, 34, 31, 34. The decrease is rather irregular; some smoothing is clearly in order. This is best done by averaging within the sample of 1000 words. For r_1, I take the average number of distinct words per block. For r_2, I take the number of words in one block not present in another and average over ten such pairs, and similarly for r_3, r_4, etc. The smoothed sequence r_n' is: 0.683, 0.487, 0.420, 0.394, 0.364, 0.336, 0.323, 0.308, 0.296, 0.283. This already begins to look more civilized, and more smoothing would likely produce even better results.

To fit a member of the family Me^{-n^t} to this sequence, I assume the r_n are all approximately normally distributed about their respective means with common variance. Given these assumptions, maximum likelihood estimates of M and t are least squares estimates, but, unfortunately, numerical methods are still needed to obtain them. There is, happily, a seat of the pants method that can be used.

The rate parameter t is reflected in the ratios r_n'/r_{n+1}' of successive terms. Setting $r_n' = Me^{-n^t}$, the natural logarithm of this ratio is $(n+1)^t - n^t$. It is increasing in n for $t > 1$ and decreasing in n for $t < 1$. Hence the same is true of the ratios. Write R_n for the observed values of these log ratios and equate to the theoretical values. This gives the equations:

$$2^t - 1^t = R_1, 3^t - 2^t = R_2, \ldots, m^t - (m-1)^t = R_{m-1}.$$

Solving successively:

$$2^t = 1 + R_1, 3^t = 2^t + R_2 = 1 + R_1 + R_2, \ldots, m^t$$
$$= 1 + R_1 + \ldots R_{m-1},$$

whence $t = \ln(1 + R_1 + \cdots + R_{m-1})/\ln m$. I apply this method to the

TABLE I

n	1	2	3	5	5	6	7	8	9	10
r'_n	0.683	0.487	0.420	0.394	0.364	0.336	0.323	0.308	0.296	0.283
Me^{-nt}		0.491	0.424	0.386	0.359	0.338	0.322	0.308	0.296	0.286

smoothed sequence, omitting the first term, and obtain $t = 0.1973444667$. The least squares estimate of M for this t is $\hat{M} = \sum_k^m e^{-kt} r'_k / \sum_k^m e^{-2kt} = 1.3359981$. The smoothed and fitting sequences are shown in Table I.

The fit is rather good (but not nearly as good if we include the first term, 0.683). The predicted number of new words in the next block of 1000 is 247, and this agrees tolerably well with the observed number, 264, especially when one considers the highly non-random distribution of words throughout the text. (Each new topic calls forth its own vocabulary.) In any case, the method is susceptible of considerable refinement.

Lagerqvist's novel contains roughly 50 000 words. The expected number of new words in the next block of 100 after 50 000 is 100×1.3359981 $\exp(-500^{0.1973444667}) = 4.4$. This does not seem wildly unrealistic. Indeed, by interrogating oneself about the expected number of new words in the next 100 after N words, for various values of N, one is led to a prior distribution of t. By contrast, I do not know how I would go about choosing a value of Hintikka's parameter α for this problem. But the real advantage of t is that it can be revised in the light of each additional sample: the modulator is modulated.

EXAMPLE 3. My colleague, Richard Edwards, has been conducting textual studies on the *Gospel According to Mark* in Greek. He did a word count for me, using blocks of 100 words each. There were 113 blocks and 2872 different words. The word counts in the first 12 blocks were as follows: 77, 60, 44, 40, 45, 45, 49, 48, 26, 29, 34, 32. Here I was unable to smooth the data. I dropped the first two observations (as both lie a long way off any smooth curve through the next eight or ten), and then fitted a curve of the family Me^{-nt}, first to blocks 3–10, and then through blocks 3–12. Since the word counts go up somewhat in blocks 11 and 12 (as compared with 9, 10), we should expect the latter curve to overestimate and the former to underestimate the word counts throughout. For the former, I find $\hat{t} = 0.1676$, $\hat{M} = 1.4065$, while for the latter, $\hat{t} = 0.1201$ and $\hat{M} = 1.2993$. The predicted number of words in the next block after 100 blocks is 16.3 in the former case and 22.9 in the latter, the observed number being 23. (The word counts for the last few blocks are suspiciously high.) As expected, the former (steeper) curve does consistently underestimate the word counts, and the latter consistently overestimates them. If we average the two, obtaining $\hat{t} = 0.1438$ and $\hat{M} = 1.3529$, we get a very satisfactory fit, especially when you consider the irregularity of the unsmoothed sample data. I give observed and expected counts in Table II, with cumulative totals.

The method developed here has two deficiencies: we haven't provided a theoretical argument for the particular family Me^{-n^t} of convex decreasing functions employed, nor have we a basis for selecting an optimal length for the blocks or samples. Work needs to be done on both problems.

Let me sketch now some extensions of the method. Having estimated the number of kinds in the population at K say, we count the number k in the sample and then add to this list the $K - k$ additional kinds which have highest

TABLE II

Block	Obs	Exp	Block	Obs	Exp	Block	Obs	Exp
1	77	–	41	24	24.9	81	19	20.8
2	60	–	42	26	24.7	82	17	20.7
3	44	49.8	43	28	24.6	83	16	20.7
4	40	44.8	44	23	24.4	84	13	20.6
5	45	41.9	45	33	24.3	85	21	20.5
6	45	40.0	46	26	24.1	86	20	20.4
7	49	38.4	47	22	24.0	87	25	20.3
8	48	37.1	48	26	23.9	88	26	20.3
9	26	36.0	49	26	23.7	89	20	20.2
10	29	35.1	50	20	23.6	90	26	20.1
Totals	326	323.1	Totals	1421	1421.8	Totals	2218	2287.8
11	34	34.3	51	20	23.5	91	19	20.1
12	32	33.6	52	18	23.4	92	29	20.0
13	29	33.0	53	19	23.3	93	21	20.0
14	32	32.4	54	25	23.2	94	16	19.9
15	30	31.9	55	24	23.1	95	10	19.9
16	26	31.4	56	13	22.9	96	20	19.8
17	30	30.9	57	21	22.8	97	17	19.8
18	17	30.5	58	25	22.7	98	20	19.7
19	26	30.1	59	16	22.6	99	21	19.6
20	33	29.7	60	19	22.5	100	23	19.6
Totals	615	640.9	Totals	1621	1651.8	Totals	2414	2486.2

TABLE II *Continued*

21	23	29.4	61	18	22.4	101	22	19.5
22	28	29.0	62	19	22.3	102	26	19.5
23	37	28.7	63	21	22.2	103	15	19.4
24	25	28.4	64	24	22.1	104	24	19.4
25	33	28.1	65	16	22.0	105	16	19.3
26	32	27.9	66	27	21.9	106	34	19.2
27	27	27.6	67	15	21.9	107	21	19.2
28	20	27.4	68	26	21.8	108	29	19.1
29	32	27.1	69	19	21.7	109	22	19.1
30	27	26.9	70	20	21.6	110	26	19.0
Totals	899	921.4	Totals	1826	1871.7	Totals	2649	2687.9
31	29	26.7	71	22	21.5	111	21	19.0
32	23	26.5	72	15	21.4	112	33	18.9
33	24	26.3	73	17	21.4	113	30	18.9
34	34	26.1	74	18	21.3			
35	19	25.9	75	22	21.2			
36	29	25.7	76	17	21.1			
37	28	25.5	77	13	21.0			
38	25	25.3	78	31	20.9			
39	30	25.2	79	16	20.9			
40	27	25.0	80	18	20.8			
Totals	1167	1179.6	Totals	2015	2083.2	Totals	2733	2735.7

a priori probability of being present in the population. There may be relations between kinds, of course, which make it more or less probable that a given kind will be present given other kinds which are present. If that is so, the initial probabilities should be revised in light of these relations and the kinds found in the sample.

It might be found, for example, in sampling a collection of wooden blocks of various shapes and colors, that color and shape are unassociated. Then, even if a general law 'All blocks of shape S have color C is consonant with a sample, its probability will remain rather low if $K - k$ is at all large, and blocks of shape S have been found.

The point to be emphasized is that we do not require 'background knowledge' in the ordinary sense to modulate our inductive inferences; we need only exploit features disclosed by the sample itself.

To compute the probability of a general law, first assign a prior distribution of t and M, and then compute their posterior distribution. This will determine a probability distribution of K. Next, for each value of K, compute the probability that kind S non-C is among the $K - k$ unsampled kinds present in the population, using the (possibly revised) chances of various unsampled kinds being present. Finally, take the weighted average of these last probabilities, the weights being the probabilities of different values of K. The method would require an inordinate amount of calculation, of course, but several practical approximations suggest themselves.

In conclusion, some remarks on infinite populations and small numbers of kinds are in order. The former are largely confined to mathematical contexts (though many populations that arise in empirical science are infinite for all practical purposes). Subject-specific arguments usually underlie probability assessments in mathematics (for examples, see the last section of Chapter 6). In other cases, one need only consider finite populations large enough that no new kinds are expected after sampling them.

Where the population is not too large and the number of possible kinds has a known and small number K, we can, in principle, apply the full-dress Bayesian method. A computer could be programmed to calculate the likelihood of all $\binom{N + K -}{K - 1}$ compositions of the population of N into K categories. On this basis, given a prior probability distribution of compositions, the posterior probability of any composition, constituent, or general law could be found. In many cases, though, it would be cheaper to sample the entire population in question than compute the relevant probabilities based on a sample.

For an alternative method of estimating the number of kinds, based on the frequencies of the frequencies with which different kinds appear in a sample, see Good (1953), who credits his leading idea to A. M. Turing. Good also gives a method for estimating the proportion of the population belonging to the sampled kinds.

BIBLIOGRAPHY

Good, I. J.: 1953, 'The Population Frequencies of Species and the Estimation of Population Parameters', *Biometrika* **40**, 237–264.
Hintikka, K. J. J.: 1966, 'A Two Dimensional Continuum of Inductive Methods', in Hintikka and Suppes (eds.), *Aspects of Inductive Logic*, North-Holland Publishing Co., Amsterdam.

PART TWO

SCIENTIFIC METHOD

CHAPTER 5

SIMPLICITY

1. INTRODUCTION

Scientific inference is thought to be hypothetical-deductive: from given facts or experimental findings we infer laws or theories from which the facts follow or which account for the facts. This is an oversimplification, though, for the facts or findings are seldom logical consequences of the explanatory theory, but merely 'agree' with the theory. Bayes' rule then enters as a more general scheme of hypothetical deduction: from given facts, to infer the most plausible theory that affords those facts highest probability.

Another constraint − simplicity − is thought to apply: among all the theories which fit the facts we are enjoined to choose the simplest. The methodologist then faces the twofold task of characterizing simplicity and then justifying preference for theories that are simplest in the indicated sense.

At first blush, the two constraints, plausibility and simplicity, seem to point in opposite directions. The simplest hypothesis is often implausible. In drawing an ace of spades from a freshly opened pack of cards, the simplest hypothesis looks to be that all fifty two cards in the deck are aces of spades.

Perhaps such considerations have led others to think of simplicity as an 'epistemic utility'. Without at all wishing to deny that simpler theories have pragmatic and aesthetic virtues, our tack will be to show why simpler theories are preferable from a strictly Bayesian point of view. While simpler theories may or may not be more plausible, they typically have higher likelihood than less simple theories which fit the data equally well. That is, *they are more confirmable by conforming data*. That is Bayesian reason enough for valuing simplicity. Before illustrating this thesis, we must of course describe the underlying conception of simplicity.

2. MEASURING SIMPLICITY

Perhaps the one firm intuition about simplicity is that we complicate a theory when we enlarge its stock of adjustable parameters.[1] (It doesn't follow, however, that a theory is simpler than another if it has fewer parameters.) In practice, when we complicate a theory in this way, the original theory is retained as a special case corresponding to a particular setting of the new parameters. E.g., a circular orbit is retained as a special case of an elliptical orbit, one of zero eccentricity. Again, when we increase the degree of a polynomial in a curve-fitting problem, the given polynomial is retained as the special case obtained by setting the leading coefficient equal to zero. We start out, then, from the principle that *special cases of a theory are simpler*.

Now the one thing we always do when we pass from a special case to a parametric extension thereof is increase the sheer number of possible experimental findings which the theory can accommodate (by the lights of a given criterion of fit). We take this feature as definatory and measure simplicity by the paucity of possible experimental findings which the theory fits. More precisely, by the *sample coverage* of a theory T for an experiment X, I mean the chance probability that the outcome of the experiment will fit the theory, a criterion of fit being presupposed. The smaller its sample coverage over the range of contemplated experiments, the simpler the theory.

A 'chance' probability distribution is one conditioned on a suitable null hypothesis of chance (e.g., an assumption of independence, randomness, or the like), constrained, perhaps, by prior background information. For present expository purposes, however, one can think of a chance distribution as a uniform distribution. In that case, sample coverage reduces to the proportion of experimental outcomes which the theory fits. (We must take Cartesian products of outcome spaces where there is more than one experiment, and the 'chance' probability distribution must reflect dependencies between the different experiments.)

How well does our measure capture pre-systematic intuitions about simplicity? Its kernel, the paucity-of-parameters criterion, is, I said, the one firm intuition about simplicity. As examples, polynomials of higher degree are more complicated, as are various well-known refinements of Mendel's laws, among them polygenes, multiple alleles, pleiotropic effects, and the introduction of additional parameters for linkage, partial manifestation, differential viability of recessives and so on.

Planetary theory (Chapter 7) provides further examples. The simplest heliostatic model, a first approximation really, is based on coplanar uniform sun-centered circles. Kepler ellipses complicate this picture, but rather minimally, since the feature of uniform circles that the radius vector sweeps out equal areas in equal times is preserved. On the other hand, Kepler's third law, by relating the motions of different planets and sharpening the empirical rule relating periods to orbital radii, imposes an extraordinarily strong kinematic constraint, drastically reducing sample coverage (i.e., the possible orbits the theory fits). There is little doubt, then, that Kepler's laws effect a net reduction of sample coverage, and hence a simplification. The same is true of Newton's theory. To be sure, Newton's orbits *look* more complicated than Kepler's, but they are just as determinate. In addition, the theory imposes a dynamic constraint; many kinematically possible Kepler orbits are excluded as dynamically unstable. Even Special Relativity may well be a simplification (Einstein clearly thought it was). One is tempted to think of Newtonian mechanics as a special case of Relativity, but, in fact, it is a *limiting case*. That is, the Einsteinian equations approximate the Newtonian when the velocity of light in vacuo approaches infinity, or, more precisely, when the speeds of the particles under consideration are small compared with the speed of light. Special Relativity, indeed, imposes an upper bound on velocities which Newtonian theory does not impose. And while the mass of a body depends on its speed in Relativity theory, it does so in a perfectly determinate way. In short, it begins to look as though Special Relativity determines more, not less, than Newtonian mechanics, and is therefore simpler in my sense. Remarking on the equality of inertial and gravitational mass, Einstein and Infeld (1938), p. 36 write:

A mystery story seems inferior if it explains strange events as accidents. It is certainly more satisfying to have the story follow a rational pattern. In exactly the same way a theory which offers an explanation for the identity of gravitational and inertial mass is superior to one which interprets this identity as accidental, provided, of course, that the two theories are equally consistent with the observed facts.

It should be clear enough from these examples and quoted passages that there is a tradition that equates the simplicity of a theory with how much it determines, with the paucity of its parameters, and the precision of its predictions. Our measure of simplicity at any rate, can hardly be considered eccentric. But a good theory must do more than square with pre-theoretic intuitions; it must systematize those intuitions and reveal unsuspected connections between them. Do our intuitions regarding generalizations, for example, have any connection with those governing parametric extensions of a theory or curve-fitting?

Consider first the concept-identification approach to generalization that has been developed by Hintikka (see Chapter 4). A *strong generalization* (or *concept*) states which of k possible *kinds* is exemplified in the population. Given that r of the k possible kinds are found in a sample, what is the simplest strong generalization compatible with the sample? Almost instinctively we answer it is that which designates all and only those kinds found in the sample. But, at first blush, this conviction seems at best remotely related to the paucity of parameters criterion, much less to sample coverage. But, as matters turn out, there is a relation, even a straightforward one.

'Fitting samples', in this context, can only mean samples which are logically compatible with the generalization in point, hence, those which exemplify a subset of the designated kinds. Clearly, the more kinds a strong generalization designates, the more samples it will fit. We conclude that the simpler of two strong generalizations is that which designates fewer kinds and that, in particular, the simplest strong generalization compatible with an observed sample is that which designates all and only those kinds exhibited in the sample. Hintikka shows that, with increasing sample size, the likelihood of this simplest law tends to infinity. That is, in the limit when sample size is infinite, the simplest strong generalization compatible with the sample becomes infinitely better supported than any alternative hypothesis. We see here an important case of the intimate connection between simplicity and support, a connection upon which I will expatiate more fully in due course. Hintikka's result was, in fact, one of the sources of my own ideas about simplicity and the grounds for preferring simpler theories.

Consider, next, generalizations of the type: All F are G. If one such generalization is stronger than another in the sense of implying it, most of us would agree the stronger generalization is simpler. On Popper's falsifiability criterion[2] this follows because the stronger generalization (logically) excludes more. Can we sharpen this criterion to one that will handle cases where neither of two generalizations implies the other? Or must we concede, with Popper, that logically incomparable hypotheses are incomparable as to their simplicity?

If we are prepared to grant that 'All F are G' is simpler than 'All F are H' when G is included in H, perhaps we might be tempted to conclude it remains simpler when H is merely more inclusive (i.e., more numerous) without necessarily including G. Analogously, 'All G are F' should count as simpler than 'All H are F' when G is more inclusive than (without necessarily including) H. Does our measure lend any support to these intuitions?

No doubt, such comparisons will depend on what we take to be the relevant experiment or question, as well as on the relevant null hypothesis or chance distribution of experimental outcomes. I readily admit my own intuitions on the comparison before us are not all that sharp, but that seems to me due entirely to the problem's not being well-posed until the relevant experiment and chance distribution are filled in. Intuitively, my tendency is to accept the cited comparisons, but that is most likely because they follow on what appear to be the most natural assumptions regarding the relevant experiment and chance distribution. Namely, they follow, if the relevant experiment is sampling the population at large and the relevant null hypothesis is that the traits in question are unassociated. An outcome will agree with the hypothesis 'All F are G', then, iff the item sampled is either an FG or a non-F, and the probability of this, on our null hypothesis, is $P(F)P(G) + [1 - P(F)] = 1 + P(F) [P(G) - 1]$, where $P(F)$ is the probability of sampling an F, etc. Let us first compare 'All F are G' to 'All F are H', where $P(G) < P(H)$. The respective sample coverages are $1 + P(F)(P(G) - 1)$, as above, and $1 + P(F)(P(H) - 1)$, and the former is smaller when $P(G) < P(H)$ – provided $P(F) \neq 0$. The same calculation shows that 'All G are F' is simpler than 'All H are F' when G is more inclusive than H (i.e., when $P(G) > P(H)$).

3. MEASURING SUPPORT

The more a theory determines, the simpler it is in my sense, and, in particular, special cases of a theory with adjustable parameters are always simpler. The rule-of-thumb that simpler theories are, *ceterus paribus*, better supported has its chief warrant in the fact that best-fitting special cases of a theory are better supported. It is natural, from a Bayesian point of view, to measure the support an experimental outcome affords a theory H with free parameters by its *average likelihood*, viz., the average of the likelihoods of its special cases against the prior distribution of its parameters:

$$(5.1) \qquad P(x/H) = \int P(x/\theta, H)\mathrm{d}p(\theta/H),$$

where θ is a free parameter (or vector of parameters). Notice, I indicate the dependence of the prior distribution on H; given another theory with the same parameters, we might assign quite different weights to given values of those parameters. Let $H:\theta = \hat{\theta}$ be the best-fitting (i.e., *maximally likely*) special case of H (that which maximizes $P(x/\hat{\theta}, H)$). Qua weighted average, $P(x/H)$ can never exceed its maximum, $P(x/\hat{\theta}, H)$, and consequently, H is never better supported (and, in practice, is always less well supported) than its best fitting special case: $P(x/H) \leqslant P(x/\hat{\theta}, H)$.

An increase in the average likelihood is criterial, on the present analysis, of a support-increasing complication of a theory. Complicating a theory (by adjunction of new parameters) will always improve (or never deprove) its accuracy but will also increase its sample coverage. These two factors have contrary affects on the average likelihood, and the trade-off, I am suggesting, is a good one, just in case the average likelihood increases. We illustrate the application of this criterion to real scientific problems in the next section of this chapter. An especially important application of the method is to regression or curve fitting problems. Raising the degree of a polynomial regression, for example, always increases sample coverage and improves fit (by a least squares or likelihood criterion of fit). The coefficients are the free parameters to be estimated from the data in this case, and so the average likelihood of a polynomial of degree $n - 1$ will be an n-fold integral. For the method to be generally applicable clearly requires an approximation to the average likelihood. An approximation based on the asymptotic normality of the likelihood function about the vector of maximum likelihood estimates of the parameters will be given here. The approximation can be used to compute by how much the maximum likelihood of an $(n + 1)$st degree polynomial regression must exceed that of an nth degree polynomial for the average likelihood to increase. I.e., the approximation can be used to determine by how much accuracy must be improved for support to increase when the degree of a polynomial regression is raised by one. The computations are carried out in Chapter 11, and the resulting Bayesian method is compared with several orthodox methods for identifying the degree of a polynomial regression.

The approximation is easily derived. Let L be the log likelihood, ln $p_x(\theta_1, \ldots, \theta_t)$, where $p_x(\theta_1, \ldots, \theta_t) = P(x/\theta_1, \ldots, \theta_t)$. (We suppress H.) This notation reminds us that the likelihood function is a function of the parameters (indexed by the observed experimental outcome). Now $p_x(\theta_1, \ldots, \theta_t)$ is asymptotically normal about the vector $(\hat{\theta}_1, \ldots, \hat{\theta}_t)$ of ML (maximum likelihood) values, with covariance matrix inverse to Fisher's information matrix with typical entry $-nE(\partial^2 L/\partial\theta_i \partial\theta_j)$, n the sample size.[3] The quadratic form Q that appears in the approximating normal density has as its coefficients the cofactors divided by the determinant of the covariance matrix. But these are just the elements of the inverse of the covariance matrix, i.e., they are the elements of the information matrix. Hence, writing I for the determinant of the information matrix, the approximating normal density can be written:

$$(5.2) \qquad (2\pi)^{-t/2} I^{1/2} \exp(-Q/2).$$

Q has a chi square distribution with t df ('df' is short throughout for 'degrees of freedom'), t being the number of parameters. Hence, writing $X^2_{\alpha,t}$ for the $100(1 - \alpha)\%$ point of the chi square distribution with t df, $Q = X^2_{\alpha,t}$ is the equation of a $100(1 - \alpha)\%$ (ellipsoidal) *Bayesian confidence set* for the parameters $\theta_1, \ldots, \theta_t$. This ellipsoid is centered at the *ML* vector and has the property that the likelihood of any included vector exceeds that of any excluded vector. Given a uniform prior distribution of the parameters, the ellipsoid would be a $100(1 - \alpha)\%$ highest density region of the posterior distribution.

The volume of the confidence ellipsoid increases with the determinant of the covariance matrix and so decreases with I, the determinant of the (inverse) information matrix. The latter, in turn, increases with the sample size n (indeed, I is of the order of n^t). Consequently, as the sample size increases, the likelihood function becomes ever more concentrated in a small neighborhood of the *ML* vector (as well as more nearly normal). This means that, in the limit, the best fitting special case of a theory becomes infinitely better supported.

The likelihood function is approximately normal in shape, but we have to apply a magnification A to the approximating normal density to make its maximum equal that of the likelihood function. (Equivalently, $1/A$ is the normalizing constant multiplication by which transforms the likelihood into a true normal density integrating to one.) Hence, we must solve $A e^0 (2\pi)^{-t/2} I^{1/2} = p_x(\hat{\theta}_1, \ldots, \hat{\theta}_t)$ for A. Now the integral of the magnified normal density is just A times the integral of the unmagnified normal density, or A, so that the factor of magnification itself should provide a good approximation to the average likelihood (the volume under the likelihood surface):

$$(5.3) \qquad A = (2\pi)^{t/2} I^{-1/2} p_x(\hat{\theta}_1, \ldots, \hat{\theta}_t) \doteq \int p_x(\theta_1, \ldots, \theta_t) d\theta \ldots d\theta_t \,,$$

writing $\doteq$ for approximate equality. N.B., 'average likelihood' means, in this context, 'average likelihood with respect to a uniform prior distribution of the parameters'. The approximation is generally not accurate where the prior is strongly non-uniform. Some illustrations of its use are given in the examples which follow.

4. EXAMPLES: MULTINOMIAL MODELS

Categorized data, such as one encounters frequently in genetics (where the categories are usually phenotypes), give rise to multinomial models. Since the distribution theory of multinomial models is well known and especially elegant, they will serve the illustrative purposes of this section well. Some slight familiarity with Mendelian genetics will be assumed.[4]

EXAMPLE 1 (genetic linkage). Two factors are 'linked' when they lie on the same chromosome. As is customary, we denote the dominant alleles by capital letters, and reserve small letters for the recessive alleles. The heterozygote can occur in either of two *phases:* coupling (*ABab*) where *A* and *B* lie on one chromosome of a homologous pair and *a, b* on the other, or repulsion (*AbaB*), where a dominant *A* is linked to a recessive *b* and vice versa. Chromosomes may break and members of a pair sometimes exchange homologous sections. As a result, the two chromosomes *Ab* and *aB* of repulsion phase may give rise to the pair *AB, ab* of coupling phase (or vice versa). The rejoined chromosomes are called *recombinants* and *crossing-over* is said to have occurred.

Let θ be the probability of crossing-over (called the 'recombination fraction'). Then the gametic output of the heterozygote *AbaB* will be *AB*, *Ab, aB*, and *ab* in the relative frequencies:

$$(5.4) \qquad \theta/2, (1 - \theta)/2, (1 - \theta)/2, \theta/2.$$

For crossing-over occurs with probability θ, and then one member of each chromosome pair is *AB* and the other is *ab*, and similarly for the non-recombinant types *Ab* and *aB*. If we cross the heterozygote to a double recessive, *abab*, a so-called 'double backcross', the four phenotypes *Ab, Ab*, *aB*, and *ab* will also be genotypes, since the recessive genes from the doubly recessive parent have nil effect. Hence, (5.4) gives the category probabilities for the double backcross, the categories being the phenotypes (and genotypes) *AB, Ab, aB, ab*. Let n be the sample size, and let n_i be the observed number of off-spring in category i, $i = 1, 2, 3, 4$. If we allow θ to range from 0 to 1, the average likelihood (with uniform prior) becomes a simple beta integral:

$$(5.5) \qquad \int_0^1 p_x(\theta)d\theta = \frac{n!\, 2^{-n}}{n_1! n_2! n_3! n_4!} \int_0^1 \theta^{n_1 + n_4}(1 - \theta)^{n_2 + n_3} d\theta,$$

$$= \frac{(n_1 + n_4)!(n_2 + n_3)!}{2^n(n + 1)n_1! n_2! n_3! n_4!}.$$

The *ML* estimate of θ is $\hat{\theta} = (n_1 + n_4)/n$, the fraction of recombinants, and its standard error is $I^{-1/2} = (\theta(1 - \theta)/n)^{1/2}$.

Mendel began with the simple special case $\theta = 0.5$ of independent assortment. Does introduction of the linkage parameter, θ, increase the theory's support? Suppose at $n = 100$ we observed 20, 30, 30, 20 in the four categories, giving $\theta = 0.4$ with standard error $I^{-1/2} = 0.048989$. The average likelihood (5.5) is 0.000127. The maximum likelihood (i.e., that of the special case $\theta = 0.4$) is 1.038×10^{-3}, and so the approximation (5.3) gives 0.000127 to six places. The likelihood of the special case $\theta = 0.5$ is 0.000139, which is larger than the average likelihood of the linkage model. The accuracy gained in this case would not be sufficient to offset the loss of simplicity; the complication of Mendel's model is support reducing for the experimental data in question. Even though the *ML* estimate 0.4 of θ is somewhat removed from $\theta = 0.5$, the sample size is too small to provide convincing evidence of linkage.

On the other hand, let us examine some of the data that actually prompted introduction of the linkage parameter. Using *Drosophila Melanogaster* T. H. Morgan backcrossed a wild-type heterozygote in repulsion phase to a double recessive having black body and vestigial wings (i.e., to *bvbv* males). For the phenotypic frequencies found and those expected on the two models, $\theta = 0.5$ and H_θ see Table I.

TABLE I

	Bvbv	*Bvbv*	*bVbv*	*bvbv*
Observed	338	1315	1552	294
Expected 0.5	874.75	874.75	874.75	874.75
Expected H_θ	314.9	1434.6	1434.6	314.9

When $\theta = 0.5$, of course, all four category probabilities are equal to ¼. The *ML* estimate of θ for Morgan's data is 0.182, which is rather more removed from 0.5, and, moreover, the sample size is much larger ($n = 3499$). One therefore expects the evidence to favor linkage rather strongly. The likelihood of the special case $\theta = 0.5$ is found to be 2.84×10^{-346}. Using (5.5), the average likelihood of the linkage model, H_θ, is found to be 1.62×10^{-12}, which is about 6×10^{333} times as great.

There is no doubt, then, that the linkage parameter is needed to account for Morgan's data. But, as the reader can see at a glance, the actual agreement between the data and the best-fitting special case $\theta = 0.182$ of H_θ, is atrocious. It is entirely possible that an even more complicated model would increase support. In fact, the data strongly suggest that neither of the two factors is segregating in accordance with the expected 1:1 ratio. An orthodox statistician would attempt to extricate these underlying assumptions of the model for separate testing; some would permit the same data to be re-used for these tests, others would not. It might be more illuminating, however, to 'correct' the observed frequencies so as to restore conformity of the individual factors with the expected 1:1 ratios, without disturbing anything else, and then see whether agreement between the data and the linkage model is restored.

To satisfy the implied constraint, the ratio of recombinants to non-recombinants must be left unchanged, so that if n_i^* are the corrected category counts, we must have $n_1^* + n_4^* = n_1 + n_4$, and hence $n_2^* + n_3^* = n_2 + n_3$. Now if the 1:1 ratios were realized, all four of the frequencies of B, V, b and v would equal 1749.5, leading to the equations:

$$n_1^* + n_2^* = 1749.5 = n_1 + n_2 + 96.5,$$

$$n_2^* + n_4^* = 1749.5 = n_2 + n_4 + 140.5,$$

$$n_3^* + n_4^* = 1749.5 = n_3 + n_4 - 96.5,$$

$$n_1^* + n_3^* = 1749.5 = n_1 + n_3 - 140.5.$$

Solving these equations subject to the cited constraints gives the corrected counts 316, 1433.5. 1433.5, 316, giving nearly perfect agreement with the data.

Before leaving the linkage example, it would be worth pausing to note an unusual feature of it. Soon after its introduction, Morgan and his students began to draw up lists of factors associated in heredity for various species of *Drosophila*. It was soon noticed that the number of these so-called 'linkage groups' always agreed with the directly observable number of chromosomes characteristic of the species, and so it was natural to hypothesize that the chromosomes were the carriers of hereditary factors or 'genes', as they came to be called. Linked factors or genes were those housed in the same chromosome, and this physical interpretation of linkage forged, if you will, a 'link' between the previously disparate sciences of genetics and cytology, a synthesis that came to be known as 'the chromosomal theory of inheritance'. Now if we confine attention to particular mating experiments, we see that the sample coverage of Mendel's theory is increased by adjunction of the linkage parameter. But if we look at classes of such mating experiments, like those which lead to the construction of linkage groups for a species, we are constrained to recognize that the cytological interpretation of linkage, because it imposes powerful new constraints – above all, that the number of linkage groups and the number of chromosomes characteristic of a species coincide – may well effect an overall reduction in sample coverage. (We have already mentioned the parallel case of Kepler's third law.) Even if this is so, the example is idiosyncratic. Other complications of Mendel's laws – polygenes, multiple alleles, partial manifestation, etc. – definitely increased the overall sample coverage of the theory, and not only its sample coverage for particular mating experiments. These were 'complications' properly so-called.

The status of a theory's parameters clearly makes all the difference. The overall sample coverage for a range of experiments will be substantially smaller if theoretical parameters are universal constants, even if, initially, there is no way to directly measure these constants. For parameter estimates from one experiment must be in concordance with those from another experiment – a powerful constraint. At the other extreme, there are parameters which must be fitted from the data anew each time the theory is applied. The linkage parameter is of intermediate type: it is a universal constant for a given pair of factors and a given species. Any over-determination of theoretical parameters by the equations of a theory will also reduce sample coverage, and the same is true, of course, of any constraints governing the parameters (e.g., inequalities, or confining them to certain regions or intervals).

EXAMPLE 2. (the $A-B-O$ blood groups). The $A-B-O$ system of blood groups was initially thought to comprise four types: A, B, AB and O. (A was later found to have subtypes.) The existence of four phenotypes demands a complication of the one-factor bi-allelic model first considered by Mendel: either we must posit more than one factor, or we must assume more than two alleles. Consider a complication of the latter kind first. The simplest, clearly, is to assume three (instead of the usual two) alleles, call them A, B, and O. These occur in the population with frequencies p, q and r, with $p + q + r = 1$. (Since r is determined from $r = 1 - p - q$, there are only two free parameters.) The O-allele of the gene functions as mere absence of A or B. Hence the genotypes AA or AO give type A, etc. In a panmictic population, an individual inherits A-alleles from both parents with probability p^2, and inherits one A-allele and one O-allele with probability $2pr$. Hence type A individuals occur with frequency $p^2 + 2pr$. The other phenotypic probabilities are easily calculated and are given in Table II.

TABLE II

	A	B	AB	O
One factor (H_1)	$p^2 + 2pr$	$q^2 + 2pr$	$2pq$	r^2
Two factor (H_2)	$s(1 - t)$	$(1 - s)t$	st	$(1 - s)(1 - t)$

The simplest form of the polygenic model, on the other hand, posits two independent factors A and B. An individual has type A iff he receives at least one dominant A gene, and similarly for type B. Let s be the proportion of persons who have at least one dominant A gene and t the proportion having at least one dominant B gene, respectively. Then an individual has type A blood with probability $s(1 - t)$, since he has at least one dominant A gene with probability s, and independently of this, recessive alleles of the B gene with probability $1 - t$. The other phenotypic probabilities for this model are also listed in Table II.

Which of these models, H_1 or H_2, is simpler? Since both have two free parameters, they cannot be compared just by counting parameters. As we will see further on, there is a rather subtle difference between them that makes H_1 slightly simpler. But it is not obvious that H_1 is simpler, and, indeed, the example was chosen partly to highlight the need for the more powerful methods of effecting simplicity comparisons developed in this chapter.

Let us compute the average likelihoods of the two models for particular data. For multinomial models, the typical entries of the information matrix are readily found to be:

$$(5.6) \qquad -nE(\partial^2 L/\partial\theta_i^2) = n \sum_{s=1}^{k} \frac{1}{p_s}(\partial p_s/\partial\theta_i)^2$$

for the diagonal entries, and

$$(5.7) \qquad -nE(\partial^2 L/\partial\theta_i\,\partial\theta_j) = n \sum_{s=1}^{k} \frac{1}{p_s}(\partial p_s/\partial\theta_i)(\partial p_s/\partial\theta_j)$$

for the off-diagonal entries, where, you recall, $L = \ln p_x(\theta_1, \ldots, \theta_t)$ is the (natural) log likelihood, and k is the number of categories.

For H_2, the two factor model, we find

$$(5.8) \qquad \hat{s} = (n_1 + n_3)/n \quad \text{and} \quad \hat{t} = (n_2 + n_3)/n$$

using $s = s(1 - t) + st$ and $t = (1 - s)t + st$. The elements of the information matrix for H_2, found from (5.6) and (5.7) above, are $I_{ss} = n/(s(1 - s))$,

$I_{tt} = n/(t(1 - t))$ and $Ist = 0$, whence,

(5.9) $I = n^2/st(1 - s)(1 - t)$.

In practice, of course, we must use the *ML* estimates of s and t in (5.9).

The elements of the information matrix for H_1 are:

$$I_{pp} = n\left(4 + \frac{4r^2}{p^2 + 2pr} + \frac{4q^2}{q^2 + 2qr} + \frac{2q}{p}\right),$$

(5.10) $$I_{qq} = n\left(4 + \frac{4p^2}{p^2 + 2pr} + \frac{4r^2}{q^2 + 2qr} + \frac{2p}{q}\right),$$

$$I_{pq} = n\left(8 - \frac{4pr}{p^2 + 2pr} - \frac{4qr}{q^2 + 2qr}\right).$$

The determinant I has no simple expression, nor do the *ML* estimates of p, q, r (which must be obtained iteratively). H_1 is a rather less tractable model than H_2.

H_2 was initially thought to be the correct model for the $A-B-O$ blood groups, but the data given below in Table III led Felix Bernstein to conclude in 1924 that H_1 is in fact the correct model. The *ML* estimates for Bernstein's data are $\hat{s} = 0.5000$, $\hat{t} = 0.2829$ for H_2 and $\hat{p} = 0.2945$ $\hat{q} = 0.1547$, $\hat{r} = 0.5508$ for H_1 (found iteratively), and I have indicated the numbers expected in the four categories on these best-fitting special cases of H_1 and H_2 in Table III.

TABLE III

	A	B	AB	O
Observed	212	103	39	148
Expected H_1	206.4	97.5	45.8	152.3
Expected H_2	180	71	71	180

It is obvious at a glance that H_1 fits Bernstein's data much better than H_2, but let us compute and compare the average likelihoods.

Let $c = \begin{pmatrix} 502 \\ 212 \quad 103 \quad 39 \quad 148 \end{pmatrix}$ be the relevant multinomial coefficient. Log $c = 269.3985$, and hence log $p_x(s, t) = \log c + \log 0.5^{251} \; 0.5^{251}$ $0.2829^{142} \; (1 - 0.2829)^{360}) = -11.5780$. Writing A for the approximation (5.3) to the average likelihood, we have log $A = \log(2\pi) - \frac{1}{2}\log I + \log$

$p_x(s, t) = -14.1280$, and so $A = 7.45 \times 10^{-15}$. The exact average likelihood is:

$$c \int_0^1 s^{251}(1-s)^{251}\, ds \int_0^1 t^{142}(1-t)^{360}\, dt$$

and the log of this intergral, using Stirling's formula to approximate $\log N!$, is found to be -14.1296, which is in tolerably good agreement with the approximate value for $\log A$ found using (5.3).

For H_1, we have $\log\ p_x(p, q) = \log\ c + \log(0.4112^{212}\ 0.1943^{103}$ $0.0911^{39}\ 0.3034^{148}) = -2.9500$, while $I = 502^2(8.4844 \times 14.8967 - 4.6680^2) = 26397319.71$, whence $\log A = \log(2\pi) - \frac{1}{2}\log I + \log p_x(p, q) = -5.8626$, so that $A = 1.3 \times 10^{-6}$, larger than the average likelihood of H_2 by a factor of about 2.3×10^8. There is no doubt, then, that Bernstein's data decisively favor H_1.

Edwards (1972), pp. 39ff, analyses the same data from a likelihood point of view. He finds that the (natural) log likelihood for the best-fitting (ML) special case of H_1 exceeds that for the best-fitting case of H_2 by nearly 20 and concludes that H_1 is decisively favored (p. 42), because, he says, two independent parameters having been fitted from the data in each case, "the two hypotheses are strictly comparable in 'simplicity' ".

Now strictly speaking, the parameters p, q of H_1 are not independent, but are subject to the constraint $p + q \leqslant 1$, and, partly for this reason,[5] H_1 is in fact simpler than H_2. The more intriguing point, however, is this: what would Edwards have said had the two models differed in simplicity by his strictly parametric criterion? Nowhere in his very elegant book does he tell us how to adjust the likelihoods of two models of a comparison for simplicity difference between them, although in the passage on p. 42 I quoted from, he does implicitly recognize that simplicity has a bearing on support. I suspect that, in practice, Edwards and other likelihood men fall back on average likelihood to compare the support of two models which differ in their numbers of free parameters. If I am right, then their treatment of the problem is essentially Bayesian, and it is hard to see how they have dispensed with prior distributions. Similar remarks apply to the problem of eliminating nuisance parameters, which a Bayesian does by integrating them out, using a prior distribution.

Orthodox statisticians, presumably, would prefer to use a chi square criterion to determine which of H_1 or H_2 is in better agreement with Bernstein's data. This test does adjust for simplicity, to the extent that one degree of freedom is lost for each parameter fitted from the data. Yet, the chi square criterion is insensitive to constraints governing the parameters and their allowed ranges, as well as to the differing sample coverages of the special cases of two theories.[6]

Before leaving the blood group example, I illustrate the computation of Bayesian confidence sets. The 99.99% point of the chi square distribution is 13.806, and hence

$$I_{pp}(p - \hat{p})^2 + 2I_{pq}(p - \hat{p})(q - \hat{q}) + I_{qq}(q - \hat{q})^2 = 13.806$$

is the 99.99% confidence ellipse for p, q. For Bernstein's data this becomes:

$$4264.21(p - 0.2945)^2 + 4685.57(p - 0.2945)(q - 0.1547)$$
$$+ 7478.12(q - 0.1547)^2 = 13.806$$

and from this I obtain the simultaneous 99.99% interval estimates:

$$p = 0.2945 \pm 0.0569 \quad \text{and} \quad q = 0.1547 \pm 0.0430.$$

I.e., one can be 99.99% sure that the proportions of the A, B and O alleles in the population from which Bernstein's sample was drawn lie in the respective intervals for p, q and $r = 1 - p - q$.

Notice, finally, that the best-fitting special case $s = 0.5$ of H_2 has average likelihood 1.33×10^{-13}, while its best-fitting special case $t = 0.2829$ has likelihood 2.64×10^{-12}, as compared with the average likelihood 7.45×10^{-15} of H_2 itself. This illustrates the slow convergence towards the infinitely greater support a best-fitting special case of a model enjoys at an infinite sample size.

5. MULTINOMIAL MODELS WITHOUT FREE PARAMETERS

Let the random vector $(X_1, \ldots, X_k)$ have the multinomial distribution with k categories and category probabilities $p_1, \ldots, p_k$:

$$(5.11) \qquad P(X_1 = x_1, \ldots, X_k = x_k) = \frac{n!}{x_1! \ldots x_{k-1}! x_k!} p_1^{x_1} \ldots p_k^{x_k}$$

where x_i is the observed number in cateogry i and $\Sigma x_i = n$, the sample size.

Then the vector $(X_1, \ldots, X_{k-1})$ has, asymptotically, a non-singular normal distribution with density

(5.12) $(2\pi)^{-(k-1)/2} \Lambda^{-1/2} e^{-Q/2}$,

where

$$Q = \sum_{i,j}^{k-1} (\Lambda_{ij}/\Lambda)(X_i - np_i)(X_j - np_j)$$

is the quadratic form whose matrix of coefficients is inverse to the covariance matrix of the multinomial with diagonal entries $np_i(1 - p_i)$ and off-diagonal entries $-np_ip_j, i \neq j$. The Λ_{ij} are thus the co-factors of the covariance matrix and Λ the determinant. Given the obvious analogy to the univariate case, Λ is called the *generalized variance* of the distribution and its square root the *generalized standard deviation*. It is well known that Q has a chi square distribution with $k - 1$ degrees of freedom. Hence

(5.13) $Q = \chi^2_{\alpha, k-1}$

is the equation of a $100(1 - q)\%$ *direct confidence ellipsoid* for the vector $(X_1, \ldots, X_{k-1})$, writing $\chi^2_{\alpha, k-1}$ for the upper $100\alpha\%$ point of the chi square distribution with $k - 1$ *df*. (More generally, the experimental outcome will fall in a $100(1 - \alpha)\%$ *direct confidence region* (or *DCR*) with probability $100(1 -\alpha)$, it being assumed that the probability of every included outcome, conditional on the model, exceeds that of any excluded outcome.) The volume of the ellipsoid $Q = c$ is given by:

(5.14) $$V = \frac{(\pi c)^{(k-1)/2} \Lambda^{1/2}}{\Gamma\left(\dfrac{k-1}{2} + 1\right)}$$

and further

(5.15) $\Lambda = n^{k-1} P$,

where n is the sample size and $P = \Pi_i p_i$ is the product of all category probabilities $p_1, \ldots, p_k$). Hence, sample coverage increases with Q, a measure of the deviation between observed and predicted values, and with P, the product of category probabilities. (Here we are using the obvious approximation to the number of sample points inside an ellipsoid by the volume of the ellipsoid.) It follows, in particular, that at a given confidence level and sample size, two multinomial models can be compared for simplicity by comparing the products of their category probabilities. The rougher the frequency counts predicted by the model, the smaller its sample coverage — as we would expect. It also follows from (5.14) and (5.15) that a model simpler than another at one sample size is simpler at every sample size.

EXAMPLE 3. Consider the uniform trinomial ($k = 3$, $P_1 = P_2 = P_3 = 1/3$ at sample size, $n = 36$. The covariance matrix is

$$\begin{pmatrix} 8 & -4 \\ -4 & 8 \end{pmatrix}$$

with determinant $\Lambda = 48$ (also given directly by (5.15) whence $\Lambda_{ii}/\Lambda = 1/6$ and $\Lambda_{ij}/\Lambda = 1/12$. The upper 5% point of the chi square distribution with $k - 1 = 2$ df is 5.99. The equation of the 95% confidence ellipse is thus

$$\frac{1}{6}(X - 12)^2 + \frac{1}{6}(X - 12)(Y - 12) + \frac{1}{6}(Y - 12)^2 = 5.99.$$

By setting $k = 3$, $c = 5.99$ and $\Lambda = 48$ in (5.14), the area of the ellipse is found to be 130.3. The exact number of included sample points is 127. Since there are

$$\binom{n + k - 1}{k - 1} = \binom{38}{2} = 703$$

sample points altogether, the exact sample coverage for $n = 36$ and $c = 5.99$ is $127/703 = 0.1806$, or roughly 18%.

Suppose next that at $n = 81$ we observe 32, 22, and 27 in categories 1, 2, 3. The covariance matrix is now

$$\begin{pmatrix} 18 & -9 \\ -9 & 18 \end{pmatrix}$$

and

$$Q = \frac{2}{27}(32 - 27)^2 + \frac{2}{27}(32 - 27)(22 - 27) + \frac{2}{27}(22 - 27)^2$$
$$= 1.85 .$$

The number of sample points is now

$$\binom{81 + 3 - 1}{3 - 1} = 3403 ,$$

the number of included sample points at $Q = 1.85$ is approximately 90.6 using (5.14) and so the sample coverage is about 2.66% where 'fitting' outcomes are those for which $Q \leqslant 1.85$. I.e., if chance prevailed, one would expect agreement this good in only 2 or 3 cases in every 100 replications of the experiment.

Consider, finally, the trinomial model with category probabilities 1/6, 1/3, 1/2. The ratio of the product of these to the product $(1/3)^3 = 1/27$ for the uniform trinomial is 3 : 4. Hence, if the two models fit the data of any experiment equally well (i.e., if their quadratic forms assume the same value at the observed outcome x), the sample coverage of the model 1/6, 1/3, 1/2 would be three-fourths that of the uniform trinomial, and so it would be slightly better supported.

(Incidentally, one might reasonably argue a better definition of 'disorderly universe' would take the order in which the sample points fall into the different categories into account. The result would be to make fat models fatter and thin models thinner. However, the calculation of sample coverage would be considerably complicated and so I have chosen, if only for expository convenience, not to follow this course.)

Returning to the general case, let H', H'' be two non-parameter multinomial models with quadratic forms Q', Q'', and assume that at the outcome x of the experiment $Q'(x) = Q''(x)$, so that the two models fit the outcome equally well in a direct confidence sense. Then, by (5.12), the likelihood ratio quickly reduces to:

$$(5.16) \qquad P(x/H''): P(x/H') = (\Lambda' : \Lambda'')^{1/2} = (P' : P'')^{1/2} ,$$

where we continue to write P's for the products of category probabilities. Consequently, *the simpler model has higher likelihood*. In a quite literal sense, then, simplicity is an ingredient of support, for simplicity and likelihood both increase with decreasing generalized standard deviation. Notice, too, that if $P' = P''$, so that the models are of equal simplicity, the natural logarithm of the likelihood ratio reduces to $0.5[Q'(x) - Q''(x)]$, so that the better fitting model then has higher likelihood.

6. INTERVAL HYPOTHESES

The previous considerations show that, of two normal hypotheses, that with the smaller variance is both simpler and better supported by 'equally agreeing' outcomes. The same must then be true if we are comparing, say, two interval hypotheses about a binomial parameter p, where agreement is measured in standard deviation units. Here I am thinking of connected (or 'one-sided') interval hypotheses, but let us see what happens when a central connected interval hypothesis is compared to the two-sided alternative.

As a special case, consider the test of a simple hypothesis, say $p = 0.5$, against the two-sided (composite) alternative $p \neq 0.5$. The former, a point hypothesis, is obviously 'simpler' in my sense as well (a sort of pre-established harmony of usage). Assume for convenience that the prior distribution of p is uniform, or, at any rate, uniform over values of $p \neq 0.5$. The average likelihood of the composite hypothesis, $p \neq 0.5$, is then

$$\int_0^1 p^x (1 - p)^{n-x} \, dp = x!(n - x)!/(n + 1)!$$

omitting the binomial coefficient, and hence the ratio of the average likelihoods (the *Bayes factor*) in favor of $p \neq 0.5$ is $2^n x!(n - x)!/(n + 1)!$, which exceeds unity iff

(6.1) $\binom{n}{x} < 2^n/(n + 1)$,

i.e., iff the observed binomial coefficient, $\binom{n}{x}$, is smaller than the average of the binomial coefficients $\binom{n}{i}$, $i = 0, 1, \ldots, n$. Consequently, the simpler hypothesis, $p = 0.5$, is better supported, except at outcomes which lie far out in the tails of the binomial distribution based on $p = 0.5$. This test has the desirable property that its power decreases systematically with sample size.

Consider next a test of an interval hypothesis, say $0.49 \leqslant p \leqslant 0.51$, against the two-sided composite alternative, $p < 0.49$ or $p > 0.51$. The 'borderline' outcome $x =$ nearest integer to $0.49n$ agrees about equally well with both hypotheses, and so we must ask: which hypothesis is favored by this outcome?

Let prior weights of t and $1-t$ be assigned the two hypotheses, and assume again that the distribution of these probability masses is uniform for each hypothesis (i.e., within each set). Our assumption is tantamount, in general, to a non-uniform distribution of p within the entire closed interval $[0, 1]$. In fact, the constant density C of p within the subinterval $[0.49, 0.51]$ is obtained from $C \int_{0.49}^{0.51} dp = t$, whence $C = t/0.02 = 50t$. Likewise, the constant density of p in the complementary subset of $[0, 1]$ is $50(1 - t)/49$. For the posterior probability of the central interval hypothesis to exceed that of its two-sided alternative then requires that $50t \int_{0.49}^{0.51} p^x(1 - p)^{n-x} dp$ exceed $50(1 - t)[\int_0^{0.49} p^x(1 - p)^{n-x} dp + \int_{0.51}^1 p^x(1 - p)^{n-x} dp]/49$. By setting $t = 0.5$, we can determine which hypothesis is better supported (and, hence, confirmed) by the 'equally agreeing' outcome $x = 0.49n$. It is found that the simpler hypothesis, $0.49 \leqslant p \leqslant 0.51$, is better supported, and the decisiveness of the result increases, as expected, with the sample size n. E.g., for $n = 2$, $x = 1$, the posterior odds are $25 : 16$, while for $n = 4$, $x = 2$, they are $2 : 1$.

Two points remain. First, a central hypothesis that is fatter than the two-sided alternative can remain better supported by an 'equally agreeing' outcome. E.g., let the central hypothesis confine p to the interval $[0.2, 0.8]$. At $n = 5$, $x = 4$, the posterior probability of the central hypothesis is

$$\frac{10}{6} \int_{0.2}^{0.8} p^4(1 - p) dp = 0.036 \, ,$$

while that of the two-sided alternative is

$$\frac{10}{4} \left[\int_0^{0.2} p^4(1 - p) dp + \int_{0.8}^1 p^4(1 - p) dp \right] = 0.029 \, .$$

This, then, is an exception to the rule of thumb that says: the simpler of two alternative hypotheses is better supported (or confirmed) by an 'equally agreeing' outcome. It is not, however, a practically important exception, as one would not ordinarily be interested in a central interval (or 'approximate null') hypothesis that was fatter than its two-sided alternative. Indeed, I know of no practically important exceptions to the stated rule. On the other hand, one's intuition is that a disjunctive hypothesis is, *ceterus paribus*, more complicated than a non-disjunctive hypothesis, and that intuition is certainly the basis of our judgment that an interval hypothesis is complicated by widening the interval. In comparing a central (connected) interval hypothesis with its two-sided (and disjunctive) alternative, however, the intuition in question cuts both ways, so to speak, and we are at a loss to say which of the two hypotheses is more complicated.

The second point worth remarking is perhaps more disquieting. It is a central tenet of Sir Karl Popper's philosophy of science that logically stronger hypotheses are, if not more *trust*worthy, more *test*worthy. And by this he seems to mean that, by testing a stronger hypothesis (which he equates with a 'more falsifiable' hypothesis), one is apt to learn more (or advance one's knowledge more). Applied to the present context, a sharp null hypothesis, $p = p_0$, should be more testworthy than an approximate null hypothesis centered at p_0 (the null value). This expectation is not realized, however, if 'apt to learn more' is equated with expected information, and, more specifically, with the concept of *discrimination information* (defined in Chapter 1). That is, a sharp null hypothesis is less discriminable from its alternative than an approximate null hypothesis, and is therefore less 'testworthy' in the indicated sense.

To illustrate, I will compute the discrimination information for $n = 10$ Bernoulli trials for the following pairs of hypotheses: (1) $H:p = 0.5$ vs $K : p \neq 0.5$, (2) $H' : p \in [0.49, 0.51]$ vs $K' : p \notin [0.49, 0.51]$, (3) $H'' : p \in [0.45, 0.55]$ vs $K'': p \notin [0.45, 0.55)$, (4) $H''': p \in [0.40, 0.60]$ vs $K''': p \notin [0.40, 0.60]$. The probabilities of x and $10\text{-}x$ are the same on all these hypotheses. I list in Table IV the outcome probabilities for $x = 0,1,2,3,4,5$.

TABLE IV

x	0	1	2	3	4	5
H	0.00098	0.00977	0.04394	0.11719	0.20508	0.24609
K	0.09091	0.09091	0.09091	0.09091	0.09091	0.09091
H'	0.00098	0.00980	0.04402	0.11723	0.20500	0.24593
K'	0.09274	0.09256	0.09187	0.09037	0.08858	0.08774
H''	0.00113	0.01065	0.04584	0.11833	0.20304	0.24204
K''	0.10088	0.09983	0.09592	0.08786	0.07845	0.07412
H'''	0.00163	0.01341	0.05136	0.12136	0.19701	0.23045
K'''	0.11323	0.11028	0.10080	0.08329	0.06438	0.05602

Notice in the Table that H is less well confirmed by $x = 3,4,5$ than is H', and H' is less well confirmed than H'', etc. Thus, H' will be more discriminable from K' than H is from K, etc. In fact, the discrimination informations (using natural logarithms) are: $I(H:K) = 0.52$, $I(H';K') = 0.54$, $I(H'';K'') = 0.61$, and $I(H''';K''') = 0.72$. The reason is clear: $x = 3,4,5$ agree closely with the central hypothesis, but these outcomes are improbable on values of p to the right of the right endpoint of the central interval, and the more so the larger this central interval is chosen.

The probabilities in question are computed using Karl Pearson's *Tables of the Incomplete Beta Function*. Using his notation (which differs from the notation for the beta function used elsewhere in this work) which central interval hypothesis to accept as strongest, given $x = 5$, should then depend both on probability and content. I resist this temptation (cf. Chapter 12 for my reasons). If $p = 0.5$ has theoretical saliency (so that a lump of prior probability mass is assigned to this value), then we have only to compare this hypothesis with H', or H'', etc., and we will find, in every case, that $p = 0.5$ is confirmed (and for the same reason that it is confirmed when tested against $p \neq 0.5$). That completely obviates the need to bring in content (which is already reflected in the probabilities) as a *deus ex machina*.

$$B_x(p,q) = \int_0^x t^{p-1}(1-t)^{q-1}\,dt\,,$$

$$B(p,q) = \int_0^1 t^{p-1}(1-t)^{q-1}\,dt\,,$$

and $I_x(p,q) = B_x(p,q)/B(p,q)$. Hence

$$\binom{10}{i}\int_{0.49}^{0.51} t^i(1-t)^{10-i}dt = \binom{10}{i}\left[\int_0^{0.51} t^i(1-t)^{10-i}dt\right.$$
$$\left. - \int_0^{0.49} t^i(1-t)^{10-i}dt\right]$$
$$= \binom{10}{i}[B(i+1,10-i+1)I_{0.51}(i+1,10-i+1)$$
$$- B(i+1,10-i+1)I_{0.49}(i+1,10-i+1)]$$
$$= [I_{0.51}(i+1,10-i+1) - I_{0.49}(i+1,10-i+1)]/11,$$

since

$$B(i+1,10-i+1) = \binom{10}{i}^{-1}\Big/11.$$

For the probabilities to sum to 1, these values must be multiplied by 50, since $\int_{0.49}^{0.51}\sum\binom{10}{i}t^i(1-t)^{10-i}dt = \int_{0.49}^{0.51}dt = 0.02 = 1/50$. The probabilities on K' are: $(50/49)[1/11 - [I_{0.51}(i+1,10-i+1) - I_{0.49}(i+1,10-i+1)]/11]$.

The probabilities on H'', K'' etc. are of exactly the same form, but differ in the values of x in $I_x(i + 1, 10 - i + 1)$ and the normalizing constants.

Suppose now that a particular value of p, like 0.5, is theoretically salient (as will often happen, e.g., in genetics). Now, in my illustration, H''' will be more strongly confirmed by the 'perfectly agreeing' outcome $x = 5$ (when $n = 10$) in a test against K''' than will H in a test against K. Because of this, there is a temptation to say that H''' ought to be 'accepted as strongest' in the light of this outcome were it not for the fact that H has more content than H'''. The decision which central interval hypothesis to accept as strongest should then depend on both probability and content.[7] I resist this temptation (cf. Chapter 12 for my reasons). If $p = 0.5$ has theoretical saliency, so that a lump of prior probability is assigned to this value, then we have only to compare this hypothesis with H', or H'' etc., and we will find in every case that $p = 0.5$ is confirmed. That completely obviates the need to bring in content (which is already reflected in the probabilities) as a *deus ex machina*.

7. OTHER APPROACHES

I was led to my own conception of simplicity as sample coverage by attempting to find what the logic of significance tests (cf. Chapter 9) has in common with the paucity-of-parameters criterion and the intuition that the simplest 'strong generalization' compatible with a sample is that which designates all and only those kinds present in the sample. My conception also has an obvious affinity with that of Sir Karl Popper, an affinity we explore in Section 4 of the next chapter. There have been a great many attempts to solve the problem of simplicity as here understood; some of the early work is critically discussed in Ackermann (1961).

Although, for a time, simplicity was given up for dead (it was relegated to the limbo of the 'merely aesthetic'), there has been, in recent years, a strong revival of interest in the problem. It would be impossible to do justice to this ferment of ideas in a reasonably short space, but mention should at least be made of the works of Good, Goodman, Friedman, Sober and Kemeny cited in the references to this chapter. Too, there is work in progress on the fascinating concept of Kolmogoroff complexity (cf. Fine (1973), Chapter 5).

The focus of much of this work is a notion of descriptive simplicity. The 'simplest theory' in the descriptive sense is, roughly speaking, that which encodes the data in the most efficient way. Some fleeting contact will be made with this notion in Chapter 8; the question of its connection, if any, with the notion of inductive simplicity at issue in this chapter is, at any rate, deserving of attention. Given its intimate connection with evidential support, the concept of simplicity developed here is at least a plausible candidate for the title 'inductive simplicity'. Evidence that scientists have tended to use 'simplicity' in something very close to my sense will be marshalled with reference to an important historical example in Chapter 7.

NOTES

[1] First proposed explicitly, I believe, by Jeffreys and Wrinch (1921).
[2] Cf. Chapter 6.
[3] Cf. Note 7, Chapter 1.
[4] A good reference for our purposes is Sinnott *et al.* (1958).
[5] Cf. Rosenkrantz (1976), Note 8.
[6] Rosenkrantz (1976), p. 184.
[7] Cf. Levi (1976), Section IV; his views are further considered in Chapter 12.

BIBLIOGRAPHY

Ackermann, R.: 1961, 'Inductive Simplicity', *Phil. Sci.* **28**, 152–161.
Birnbaum, A.: 1969, 'Concepts of Statistical Evidence', *Philosophy, Science and Method* (S. Morgenbesser, P. Suppes, M. White, eds.), St. Martin's Press, New York.
Edwards, A. W. F.: 1972, *Likelihood*, Cambridge University Press, Cambridge.
Einstein, A. and Infeld, L.: 1938, *The Evolution of Physics*, Simon and Schuster, New York.
Friedman, K. S.: 1972, 'Empirical Simplicity as Testability', *Brit. J. Phil. Sci.* **23**, 25–33.
Good, I. J.: 1968, 'Corroboration, Explanation, Evolving Probability, Simplicity, and a Sharpened Razor', *Brit. J. Phil. Sci.* **19**, 123–43.
Good, I. J.: 1969, 'A Subjective Evaluation of Bode's Law and an "Objective" Test for Approximate Numerical Rationality', *J. Amer. Stat. Assoc.* **64**, 23–66.
Goodman, N.: 1959, 'Recent Developments in the Theory of Simplicity', *Phil. and Phenom. Research* **19**, 429–446.
Goodman, N.: 1961, 'Safety, Strength and Simplicity', *Phil. Sci.* **28**, 150–151.
Hintikka, K. J. J.: 1967, 'Induction by Elimination and Induction by Enumeration', *The Problem of Inductive Logic* (I. Lakatos, ed.), North-Holland Publ. Co., Amsterdam.
Jeffreys, H. and Wrinch, D.: 1921, 'On Certain Fundamental Principles of Scientific Inquiry', *Phil. Mag.* **42**, 369–390.
Kemeny, J. G.: 1953, 'The Use of Simplicity in Induction', *Phil. Rev.* **62**, 391–408.
Kemeny, J. G.: 1955, 'Two Measures of Simplicity', *J. Phil.* **52**, 722–733.
Popper, K. R.: 1959, *Logic of Scientific Discovery*, Hutchinson, London.
Post, H. R.: 1960, 'Simplicity in Scientific Theories', *Brit. J. Phil. Sci.* **11**, 32–41.

Rosenkrantz, R. D.: 1976, 'Simplicity', *Foundations of Probability Theory, Statistical Inference, and Statistical Theories of Science* (W. A. Harper and C. K. Hooker, eds.), D. Reidel, Dordrecht, Vol. 1, 167–203.
Smith, C. A. B.: 1969, *Biomathematics*, Vol. 2, Griffin, London.
Sober, E.: 1975, *Simplicity*, Oxford University Press, Oxford.
Sinnott, E. W., Dunn, L. C., and Dobzhansky, T.: 1958, *Principles of Genetics*, McGraw-Hill, New York.

CHAPTER 6

BAYES AND POPPER

1. INTRODUCTION

There has been little direct confrontation between Bayesians and Popperians. No doubt Popper views the Bayesian approach as a species of inductivism and therefore not to be countenanced. But while Bayesians evaluate hypotheses primarily in terms of their probability, their position rests on no obscure 'principle of induction', but on Bayes' rule. For a Bayesian, 'learning from experience' can only mean modifying prior probabilities by conditionalizing on observed experimental outcomes. On the other hand, the implications of Bayes' rule for scientific method often have a distinctly Popperian ring, fostering the suspicion that many of the differences between the two schools are, at worst, merely verbal, and, at best, but differences of emphasis. We shall have something to say about this, but, in a more positive vein, we wish to explore here the extent to which Bayesian and Popperian viewpoints can be reconciled.

2. CONFIRMATION VS CORROBORATION

Perhaps no single issue has generated more controversy than the alleged distinctness of corroboration and confirmation. For a Bayesian, 'confirmation' means 'raising the probability of'; for a Popperian, 'corroboration' (a term Popper coined to avoid confusion with probabilistic notions of confirmation) means 'withstanding severe tests'. But what are 'severe tests'? Their primary attribute seems to be this.[1] Let x be an outcome of the test or experiment that agrees with the hypothesis H. Then the test of H is severe if $P(x)$ is small. The magnitude of the deflection of light by a large body predicted by relativity theory was double the Newtonian defection, and consequently, the Eddington expedition of 1919 to observe the solar eclipse of that year constituted a severe test of relativity. As this example rightly suggests, severity is as much a property of the theory under test as it is of the experiment designed to test it. The aim of experimentation, from this perspective, is to make it virtually impossible for the experimental outcome to fit the theory unless the theory is true (or, at any rate, an adequate representation of the experiment in hand).[2] Now, as everyone knows, it is a consequence of Bayes' rule that hypotheses are confirmed by their consequences (or, more generally, by outcomes which they afford a high probability), and the more so as the outcomes in question are otherwise surprising or improbable. In the extreme case where x is a logical consequence of H, Bayes' rule reduces to $P(H/x) = P(H): P(x)$, which expresses the posterior probability of H as the ratio of its prior probability to the outcome probability $P(x)$. In short, Bayes' rule is not only compatible with, but actually rationalizes the primary attribute of corroboration.

If that were the only attribute of corroboration, there would be no reason to distinguish it from confirmation. But, of course, there is more to it. Above all, there is Popper's insistence that high corroboration must not be equated with high probability. But then, equally, high confirmation cannot be equated with high probability. For if the initial probability of an hypothesis is low, say 0.001, then its probability may be greatly increased, say to 0.5, without making its final probability high. Or, at the other extreme, tautologies are highly probable, but not at all confirmable.

Moreover, while Popper maintains that high corroboration only makes an hypothesis *test*worthy (never *trust*worthy), that heroic line is difficult to maintain in practice, or, more precisely, in spelling out the relation between theory and practice. Mendelian genetics is better corroborated than the blending theory, and of course that can be expected to have a bearing on eugenic proposals, genetic counseling, breeding and agricultural programs, nature vs nurture disputes, and so on. It is almost impossible to resist the conclusion that because the particulate theory is better corroborated, it is rational to premiss it in theoretical derivations and practical decisions. But to act as though that theory were true is to assign it a higher probability than any of its rivals (it is to bet on its being the best available approximation to the truth in this domain).[3]

Even if we grant the contention of the last paragraph, it would be premature to conclude that confirmation and corroboration are identical. For there are passages in which Popper insists that a severe test must represent our 'sincere attempts to overthrow' the tested theory.[4] Indeed, the Popperian prescription for the growth of knowledge is to attempt, with all the technical and intellectual resources at our command, to overthrow well-corroborated theories and to replace them by ever more specific, more precise and comprehensive theories, by ever 'bolder conjectures'. The sincerity requirement is surely vague and even smacks of the psychologism Sir Karl is normally at pains to disparage. To be sure, a scientist's 'set' has much to do with what he observes (or, if you prefer, to the construction he places on his findings). Popper contends that 'confirmations' of the most absurd theories are easy to come by if only you look for them. But, equally, as he admits, disconfirmations of a promising theory are also easy to come by if you display sufficient unwillingness to refine or amend a theory to handle exceptions. Early critics of Mendelism (and I am thinking here particularly of Weldon) were often content to find cases in which the 3 : 1 ratio did not apply, or in which complete dominance was absent. Nevertheless, while an overly critical or an overly receptive stance may lead to facile 'refutations' or 'confirmations' that lack real force, it is crucially important, from a Bayesian point of view, to distinguish the effect of intentions on what sorts of tests are performed from their lack of effect on the import of the results. Once we agree that the bearing of an experiment is registered in the probability changes it effects, and therefore that the likelihood function conveys the entire import of the experiment, we see that the experimenter's intentions (which have no affect on the likelihood function) have no bearing on the interpretation of his results.[5] As I. J. Good observes, a statistician might stop sampling because his pet hypothesis was ahead of the game.[6] "This might cause you to distrust the statistician", he says, "but if you believe his observations, this distrust would be immaterial".

Perhaps the view that a theory can be genuinely corroborated only if it is exposed to risk of falsification is the real substance of the sincerity requirement. And there is truth in the claim that the more a theory risks falsification (i.e., the fewer experimental outcomes it fits) the more it is confirmed by an outcome that it does fit. Still, this requirement is no more than a rule-of-thumb. The exceptions are obvious and involve the indirect confirmation of a law by disconfirmation of its alternatives. We may know, for example, that either all A are B or no A are B. Sampling B's cannot possibly falsify the former alternative, but, should a known B prove to be an A, that would eliminate the latter alternative, and so, not merely confirm, but establish the former. This shows, too, that confirmation depends on what are taken to be the alternative hypotheses — as it should.

By now the relation between confirmation and corroboration should begin to exhibit a familiar pattern. Corroboration is a qualitative notion that embodies a number of plausible intuitive canons, and the latter either follow (usually in sharpened form) from Bayesian confirmation theory or are seen to be exceptionable rules-of-thumb. Popper, of course, expressly doubts whether there can be any benefit in attempting to quantize the relevant canons or formally explicate the intuitive notion of corroboration.[7] But that line has always seemed to me to go against the grain of Popper's advice to seek 'bolder' and more precise theories wherever possible. Bayesian confirmation theory is 'bolder', not only for being quantitative, but also for having more radical implications. I have already alluded to the principle that the entire import of an experiment is conveyed by the likelihood function at the observed outcome. This implies, among other things, the irrelevance of the experimenter's intentions when to stop sampling, of his hopes of confirming (or of disconfirming) the theory, of outcomes that were not but might have been observed, and of whether the theory was formulated in advance or suggested by the observations themselves. Orthodox (non-Bayesian) statisticians have found this to be strong medicine indeed! Bayesian resolutions of the paradoxes of confirmation also deserve mention in this context.

Popper neatly evades the paradoxes, for on his view, a 'positive case' of a law will only confirm it when its observation results from a *test* (viz., a sincere attempt to refute it). Investigating colors of emeralds before time t need not falsify Goodman's grue hypothesis, be it ever so false, and there is therefore no reason why finding green emeralds before t should be expected to corroborate the hypothesis that all emeralds are grue. The Bayesian solution is, if anything, more radical (and I would add, more insightful), for it shows, not merely that a law need not be confirmed by its positive cases, but that it may actually be disconfirmed (see Chapter 2).

Much more can be said about both paradoxes of confirmation from a Bayesian point of view than is said in Chapter 2, but perhaps enough is said there to suggest the greater 'resolving power' of a probabilistic analysis.

Incidentally, the belief that laws are confirmed by their 'positive cases' has been an integral part of inductivism for a long time, perhaps the cornerstone. Bayesian analysis shows that this once apparent truism is in fact a rather clear-cut error, and that should further discourage any facile equation of Bayesianism with inductivism. Popper has defended his thesis that highly probable theories are not the goal of science in part by appealing to the ease with which one can find confirming cases of a theory if one looks for them.[8] I agree that it is easy to find conforming cases, viz. observations consistent with a theory or general law. Red herrings and white crows, however, while they are assuredly conforming cases of the raven hypothesis, are no more confirming cases than they are corroborating cases. As we have seen, they may even be disconfirming cases. Popper's argument is devoid of force, then, when directed at the Bayesian position; it may show there is something wrong with Hempel's definition of a 'positive case', but it has no tendency to show that theories cannot be appraised in terms of probability.

3. DEMARCATION

The likelihood principle implies, as already mentioned, the irrelevance of predesignation, of whether an hypothesis was thought of beforehand or was introduced to explain known effects. Bayesians deny that any additional force attaches to agreeing outcomes predicted in advance (though they would not deny that the fertility of a theory shows itself in the novel experiments it suggests). But a belief in the peculiar virtue of prediction is a recurrent theme in Popper's writings.[9] It is a fundamental part of his proposed demarcation of science from metaphysics. As I understand it, that proposal consists of two strands, a logical and a methodological, as it were. The *logical requirement* is that the theory logically exclude some possible state of affairs (in order to be accounted 'scientific'), or that it have 'potential falsifiers'. The *methodological requirement* is that proponents of the theory be willing to countenance at least some of the potential falsifiers as sufficient to reject the theory. In a discussion of the scientific credentials of psychoanalysis Popper writes:

'Clinical observations', like all other observations, are *interpretations in the light of theories* ... and for this reason alone they are apt to seem to support those theories in the light of which they were interpreted. But real support can be obtained only from observations undertaken as tests (by 'attempted refutations'); and for this purpose *criteria of refutation* have to be laid down beforehand: it must be agreed which observable situations, if actually observed, mean that the theory is refuted (Popper, 1963, p. 38, Note 3).

In other places Popper expresses this as a demand that one specify 'crucial experiments' by which to discriminate a new theory from its rivals in the field. Posed in this milder form, it reads like good counsel, but it is also good Bayesian counsel. In Bayesian terms, a *decisive test* of H against K is one for which the expected weight of evidence (i.e., the expected log likelihood ratio) for H against K is high conditional on H, and *vice versa*. But high likelihood ratios are enough to insure objectivity! Their evidential weight is in nowise augmented by use of a predesignated rejection rule; they speak for themselves. Which theories our practical decisions and theoretical derivations are predicated upon must be decided on the merits of the case at hand by using one's evaluations of the probabilities and the consequences of error to guide one. No purpose is served by attempting to lay down conventions which state conditions under which the relevant scientific community should 'accept' or 'reject' a theory. For imagine that the author of a new theory specifies results he would count as decisive against his theory, but it turned out that those negative findings could easily be explained without sacrificing plausibility or the theory (the supposed prediction might even turn out to have been fallaciously derived, or to rest on assumptions that can be independently infirmed). In that event, his declaration would be entirely without affect on his colleagues, and rightly so.[10]

Popper, too, seems to sense this, for he grants that one ought sometimes to adhere tenaciously to a theory beset by discrepant data: "He who gives up his theory too easily in the face of apparent refutations will never discover the possibilities inherent in his theory" (Schilpp, 1974, p. 984). He even gives two beautiful illustrations of cases where it was rational to disbelieve the observations rather than the theory they appeared to refute.[11] More to the point, Popper adds to his predesignationist stricture the qualification that no 'basic statement' can falsify a theory unless it is accompanied by a well-corroborated 'falsifying hypothesis'.[12] But that certainly makes it sound like a matter of comparing two probabilities, that of the falsifying hypothesis and that of the theory it rules out. Indeed, upon reading the fine print, one sees that Sir Karl's methodological proposal comes to no more than this, and I quote him:

Propose theories which can be criticized. Think about possible decisive falsifying experiments — crucial experiments. But do not give up your theories too easily — not, at any rate, before you have critically examined your criticism (Schilpp, 984).

Now the suggestion that scientists should be critical is hardly earth-shaking, nor does it really 'demarcate' science from related disciplines. Philosophers and methodologists, too, should be critical, should think up cases that discriminate their proposals from others, and respond to effective criticism by amending their views. (Good advice, however, is easier to give than to follow.)

We turn next to Popper's logical proposal. Theories that fit every possible outcome of an experiment cannot explain any outcome (for to explain why x occurred is, in part, to explain why other possible outcomes did not occur). Still, one can quarrel with Popper's formulation of this intuition, for it requires that a theory *logically* exclude some possible state of affairs, and probabilistic theories, like statistical mechanics and Mendelian genetics, do not logically exclude anything. But this defect is easily set right by introducing a suitable (probabilistic) criterion of fit.

My discomfiture lies rather in the fact that many of the most admired scientific theories — among them, quantum, statistical, and Newtonian mechanics, particulate inheritance, neo-Darwinism — exclude very little. To be sure, particular specifications of these abstract schemas applied to particular classes of experiments may fit a relatively small proportion of possible experimental outcomes. This is clearly true, for example, of the one-factor bi-allelic model with complete dominance and independently assorting unit characters Mendel applied to various of his experiments with garden peas. But if we allowed all of the subsequent complications of Mendel's simple model — multiple alleles, linked and interacting systems of genes, polygenes, pleiotropic effects, etc. — in any single application, the theory would fit any outcome. Similarly, Newton's second law is the core of classical mechanics, but if we are allowed free choice in specifying the force functions, this theory, too, would fit very nearly any system of moving and interacting mass points imaginable. There is reason, then, to question Popper's claim that 'degrees of empirical content' are appropriately measured by how much a theory excludes. My counter suggestion is that the most admired theories are admired for being *high-powered:* in terms of a very sparse conceptual apparatus, they can be specialized to account for a great diversity of phenomena.[13]

But so much hinges, obviously, on what one means by 'theory'. I call the high-level theories in question here 'leading principles', 'blueprints', or 'theoretical schemas'. The terms 'paradigm' (Kuhn) and 'research programme' (Popper and Lakatos) have been used in related senses. Popper, I think, regards theoretical schemas as 'metaphysical' (or nearly so), and therefore would not have them in mind when he speaks of 'subjecting our theories to the fiercest struggle for survival'. That is why (in his *Autobiography* in Schilpp) Popper can tell us that neo-Darwinism is 'metaphysical' and then go on to list testable consequences of evolutionary theory. For the consequences in question are really consequences of particular models of particular evolutionary phenomena, like balanced polymorphism, or the rate at which heterozygotes disappear from an inbreeding population. Little substantive disagreement between the three mentioned authors remains, so far as I can see, when their terminological differences are set aside. All agree that theoretical schemas are not initially, but may become, objects of experimental test. If persistent efforts to fit a non-vacuous particulate model to the $A-B-O$ blood group data met with failure, for example, geneticists would in time come seriously to entertain the hypothesis that not all inheritance is particulate. In attacking such a 'puzzle' (in Kuhn's sense), the geneticist begins with the simplest particulate models that appear to be consistent with his data, only complicating them as the need arises. Now the need here is for a criterion of *progress:* when does the improved accuracy resulting from a complication of a theory off-set the loss of simplicity? Repeated complications of Ptolemaic astronomy did not produce appreciably greater accuracy, and so the development of the theory came to be viewed as 'degenerative' ('epicyclic' has even come to be used in that sense). But repeated complications of Mendel's original model did radically improve the theory's goodness-of-fit over various classes of experiment. We have already seen (Chapter 5) a Bayesian analysis of support for models with many parameters can be used to define a *non-arbitrary* rate-of-exchange between simplicity and accuracy, leading to the required criterion of progress.

4. SIMPLICITY

Popper equates simplicity with empirical content or falsifiability: the more a theory excludes (the more 'potential falsifiers' it has), the simpler it is. His formulation is open, however, to the familiar objection (the so-called 'tacking paradox') that, by conjoining extraneous hypotheses to a theory we increase its content, but it is far from clear that such arbitrary embroidering of a theory simplifies it. Too, there is the objection that pairs of theories incomparable in content may yet be comparable in simplicity. But both objections may well be 'scholastic', for a rather minor adaptation of Popper's conception appears to handle both: viz., one *relativizes* simplicity to an experiment and one introduces a probabilistic version of content,[14] as in Chapter 5.

Most of the Popperian machinery is readily adapted to the framework of Chapter 5. Having fixed upon a confidence level $1 - \alpha$, all outcomes outside the $100(1 - \alpha)\%$ DCR may be regarded as 'potential falsifiers', and this leads straightaway to a probabilistic extension of Popper's notion of content which is then applicable to probability models of experiments, like the multinomial models, linear models, Markov models, etc. encountered in genetics, econometrics, mathematical psychology, and elsewhere. We can also measure and compare the severity of two tests of the same theory quantitatively simply by comparing the sample coverages of the theory for the two experiments or tests at a given confidence level. And, in an extended sense, the *dimension* of a theory (relative to an experiment and confidence level) is the smallest sample size at which its sample coverage becomes less than 100%.

Major benefits flow from this minor re-tooling of Popper's conception. Most important of all, simplicity comparisons can always be effected, whether we are dealing with deterministic or probabilistic theories, for the question which of two theories has smaller sample coverage can always be decided after a finite amount of calculation. Too, the tacking paradox is blunted to at least this extent: if the conjoined hypothesis is truly extraneous or irrelevant, it will have no effect on the sample coverage of the theory for a particular experiment. Thus, conjoining an hypothesis about the origin of quasars would presumably have no effect on the proportion of possible phenotypic frequency counts fitted by a Mendelian model. On the other hand, conjoining hypotheses will reduce the prior probability of the theory. We emphasize, finally, that these and other benefits of our reformulation are purchased at no essential loss of generality, for we can just as easily relativize to a finite class of experiments as to a single experiment, and so compare two theories over the intended range of their application.

Popper has argued that simpler theories are preferable because they are more falsifiable, that is, there is the *practical* consideration that they can be eliminated most quickly when false. Hence, knowledge can be expected to advance most rapidly by testing simplest or 'boldest' conjectures.

Now there is something to this, for simpler theories give sharper predictions, generally speaking, and if false, deviations from their sharper predictions of detectable magnitude can be expected to occur with higher frequency. But it is another shortcoming of Popper's original formulation that the connection between ease-of-elimination and simplicity only holds if we relativize simplicity to an experiment. In Popper's semantical sense, 'falsifiable' refers to content, to how much a theory *logically* excludes, while, in the methodological sense here in question, 'falsifiability' refers to the (average) number of observations required to refute a theory when that theory is false. Consider now problems (I call them 'deterministic') with the property that the outcome of a relevant test or experiment can be predicted with certainty once the state of nature is known. 'Twenty questions' is a paradigm, and a simplified version of this would be the problem of locating a particular square on a checkerboard or a particular number between 1 and N which you have in mind. The Popperian strategy is to ask maximally specific questions (e.g., 'Are you thinking of 32?'), but, in fact, informational analysis of deterministic problems (Chapter 1) easily shows the optimal strategy to involve posing questions which split the remaining possibilities as evenly as possible, or to run experiments whose outcomes are maximally uncertain.

Notice, now, in the 'odd ball' problem of Chapter 1, the hypothesis that the 'odd ball' is one of four can be refuted (when false) in a single weighing, by weighing four of the remaining balls against the other four, for then the pan *must* fail to balance. But no single weighing can refute the more specific hypothesis that the odd ball is ball #1, be it ever so false, for whatever combination of balls you weigh, the pans may balance, an outcome compatible with the hypothesis. Hence, while the hypothesis that singles out a single ball has higher content in Popper's sense, it has fewer 'potential falsifiers' for the experiment at hand. I take this to be yet another point in favor of relativizing our considerations to an experiment. But the example should also make us chary of concluding post-haste that there is any single strategy that will lead us most rapidly to the truth whatever the problem-context might be.

In any event, whatever the practical benefits of testing or 'accepting' simpler theories might be, there is, from a Bayesian point of view, a far more immediate and more compelling reason for preferring simpler theories. As seen in Chapter 5, *they are generally better supported*.

5. CONVENTIONALIST STRATAGEMS

Theories can be saved by explaining away discrepant data, by altering operational definitions, by invoking auxiliary hypotheses, or by complicating them through introduction of new parameters. The last of these tacts will concern us in this section. But it should be noted that auxiliary hypotheses are often best thought of under this head. When we posit an unseen body to explain the perturbations of a planet or star, we are, in effect, adding the orbital elements of the unseen body as new parameters, and, likewise, when we attempt to account for failure of the expected 3:1 ratio in a genetical experiment by claiming recessives are less viable, the effect is to introduce a parameter of differential viability.

Popper contends that complications of a theory are justifiable only when they increase the theory's content or testability.[15] But complicating a theory with additional parameters always increases sample coverage (reduces the number of 'potential falsifiers'), and in that straightforward sense, at least, reduces the refutability of the theory by the experiment at hand. And yet, complicating a theory to accommodate negative experimental findings may be perfectly justified for being support-increasing (Chapter 5).

There are well-known cases in genetics where dominance is absent, where the heterozygote is phenotypically distinguishable from both homozygotes. Thus, the progeny from mating a red to a white snapdragon are pink. To allow such 'intermediates' is to give up a very strong prediction of the theory, and, since no novel effects become predictable thereby, it is hard to see how that manoeuvre makes the theory more testable. All that can be said is that the theory remains testable and, in particular, experimentally discriminable

from its chief rival, the blending theory. For it predicts that the progeny from
two pink snapdragons will be red, pink, and white in a 1: 2: 1 ratio, so that,
unlike the blending theory, the particulate theory does not have the awkward
consequence (really the decisive objection to the blending theory) that
variation in any character would eventually disappear in a random mating
population. Yet, if we weaken the Popperian condition to require only that
the theory remain testable *simpliciter*, we have a very weak condition – one
that is satisfied by many an 'epicyclic' theory. In fine, I don't see how one
can pass on a complication of a theory without looking at the data which
prompted that complication in detail, and seeing how great an improvement
in accuracy is purchased at how great an increase of overall sample coverage.
Other factors, most especially, the plausibility of an auxiliary hypothesis, or
the interpretability of a new parameter, are no doubt important. But the
really decisive consideration, it seems to me, is whether the proposed
complication of the theory is support increasing or support reducing.

Faced with non-conforming data, one can go on complicating a theory
indefinitely, each time improving its accuracy somewhat at a loss of some
simplicity. It is the observed rate-of-exchange, whether observed sample
coverage increases or decreases, that spells the final fate of a research
programme. A theoretical scheme may be discredited to the satisfaction of all
but a few diehards in time, if successive complications result in no appreciable
improvement in accuracy, and so fail to increase average likelihood. But the
force of such repeated failure is cumulative; there is no one straw that finally
breaks the camel's back.

At the same time, our analysis shows how an upstart theory can be quite
properly considered more promising than a well-entrenched and more
accurate rival, as when it is able to resolve, at least qualitatively, an anomaly
of the older theory. The point, of course, is that the upstart theory may be
inaccurate and yet be improbably accurate.[16] It may be so much simpler
that, even though less accurate, its average likelihood is larger than that of the
older theory.

6. THE PROBABILITY–IMPROBABILITY CONUNDRUM

We come at last to the issue which appears to divide Bayesians and Popperians
most sharply. On whatever other points they may differ, all Bayesians agree
that posterior probability is the primary yardstick by which hypotheses are to
be compared. Popper, in stark contrast to this, contends that the aim of

science is to attain improbable theories, or that, of two theories concordant with the facts, the less probable theory is to be preferred. This certainly has a paradoxical ring, and yet there is a simple and seemingly compelling argument for the Popperian view. Science strives after an ever more comprehensive and more detailed account of phenomena. (It is even a convention of science to assert the strongest result one is in a position to assert.) But stronger or more specific theories, by the probability calculus alone, must be accorded *lower* probability. In particular, simpler theories are less, not more, probable.

This argument is indeed incontrovertible, and yet an aroma of paradox hovers about the conclusion. The reason is not far to seek. First, from the fact that a proposition must have lower probability than its consequences, it does not follow that simpler theories or laws are less probable – at least not on our account. For there are logically incomparable pairs comparable in simplicity. Hintikka's 'strong generalizations' (Chapter 4) are cases in point. A law which designates fewer kinds is simpler than, but does not logically imply, a law which designates a proper superset of those kinds. And there is normally a saturation point beyond which it becomes less plausible to posit additional kinds (as in estimating the number of distinct words in a book, especially a children's book). Likewise, 'All F are G' is simpler than 'All F are H' when the former projects a rarer trait, even when neither G nor H is a subset of the other. These examples show clearly that simplicity and probability are not always inversely related (and they are clearly not always directly related either).

Even where a simpler theory or law is a special case of another, the claim that it is less probable, though strictly true, continues to have a paradoxical ring, and again, I think, for good reason. In the linkage example of Chapter 5, for instance, we are not actually interested in comparing H_θ with its special case $\theta = 0.5$, but in comparing the latter with $\theta < 0.5$. And this pair of hypotheses is mutually exclusive (the two even point to distinct physical possibilities – the two genes lying on the same or on distinct chromosomes). Nevertheless, $\theta = 0.5$ is still *effectively* a special case (and, in any event, simpler than) $\theta < 0.5$. For any sample fitted by $\theta = 0.5$ will be fitted as well by $\theta < 0.5$, for a value of θ sufficiently close to, but distinct from, 0.5.

We tend to assume this analysis implicitly in practice. It makes good sense to ask, for example, whether an orbit is a circle or an ellipse, but for the question to have point, the 'or' must be exclusive and 'ellipse' construed as 'non-degenerate ellipse (of positive eccentricity)'. At any rate, we can always append the appropriate non-triviality clause explicitly without essentially affecting the relevant simplicity comparisons (e.g., by requiring that the leading coefficient of a polynomial be non-vanishing, we logically separate polynomial hypotheses of different degrees, but an nth degree polynomial, so construed, remains effectively a special case of an $(n + 1)$st degree polynomial by allowing the leading coefficient of the latter to approach zero). There is really no sound basis, then, for concluding that simpler hypotheses are necessarily less probable, and I think the linguistic conventions cited here completely explain the paradoxicality of any such claim. The issue that *appears* to divide Bayesians and Popperians most sharply thus dissolves. It is not merely consistent to believe that scientists ought to seek theories of minimal overall sample coverage and, at the same time, believe it appropriate to evaluate theories in terms of their probabilities. For theories of minimal sample coverage are preferable just because they are capable of attaining the highest posterior probabilities.

I do not claim, of course, to have disposed of all the arguments against appraising theories probabilistically. Carnap's system of inductive logic assigns zero probability to a generalization in an infinite domain on the basis of any *finite* sample. This was certainly grist for Popper's mill. General laws have high content, and so, on his view, they should have low probability. But for reasons I do not entirely fathom, Popper *appears* to think that general laws *must* have zero probability.[17] If that could be shown, it would certainly demolish the Bayesian programme of evaluating theories and laws in terms of their probabilities. But I know of no sound arguments to this effect.

Indeed, there are countervailing considerations: the requirement of strict coherence is violated by assignment of zero probabilities to propositions that are not logically impossible.[18] The only *prima facie* difficulty for Bayesians here is that the 'informationless' prior (based on no data) for Hintikka's problem, which is based on the assumption that individuals are assorted among kinds at random, does assign zero probability to every strong generalization in an infinite universe, save the 'trivial' one which asserts the non-emptiness of all k kinds.

The dilemma is resolved simply by assigning the non-trivial strong generalizations *infinitesimal* probabilities. To the inevitable objection that such an assignment is *ad hoc*, there is a rather compelling reply. For it is natural to require that the probability of a law in an infinite universe (based, perhaps, on an infinite sample) should be the limit of its probabilities in finite universes as their size approaches infinity. Let H^* be the law which designates c of the k kinds, $c \leqslant k$, and assume that the sample x exemplifies all and only those c kinds. By the methods of Hintikka (1967), we find that the posterior probability of H^* is approximately:

$$P(H^*/x) \doteq \frac{1}{1 + (k-c)\left(\dfrac{c+1}{c}\right)^{N-n} + \cdots + \left(\dfrac{k-c}{k-c-1}\right)\left(\dfrac{k-1}{c}\right)^{N-n} + \left(\dfrac{k}{c}\right)^{N-n}},$$

where N is the population size and n the sample size. Let both n and N approach infinity. If $N - n$ also becomes infinite, so does the denominator above, and so $P(H^*/x)$ is zero (or infinitesimal). But if N and n tend to infinity in such a way that $N - n$ remains bounded, the denominator above also remains bounded, and consequently, in the limit, $P(H^*/x)$ is finite. Since its posterior probability can be finite, H^* must have had a non-zero prior probability. Hence, we are *forced* to assign infinitesimal (rather than zero) probabilities to the non-trivial strong generalizations.

Personally, I find nothing intuitively upsetting in these results. To assign zero prior probability would preclude learning from experience; to assign a strong generalization positive prior probability in an infinite disorderly universe (i.e., on the basis of no evidence) would violate the axioms of probability, if one accepts the rather compelling representation of ignorance by the assumption that individuals are assorted among kinds at random. Having assigned infinitesimal prior probability, only infinitesimally probable evidence could raise the prior probability to a finite positive value. This, too, I find reasonable. After all, we are dealing here with very bare data; we are told only which kinds are exemplified in the sample, but we are not told even the frequencies with which the different kinds are exemplified. Mathematicians, to be sure, repose high degrees of credence in various unproved number theoretic conjectures, but only on the basis of more sophisticated evidence. E.g., nobody takes the Goldbach conjecture seriously because they have found many even numbers representable as Goldbach sums (viz. sums of two odd primes). Our confidence would increase, though, if we found that the number of representations of $2n$ as a Goldbach sum was increasing with n, or, better still, if we found (as really appears to be true) that the number of Goldbach sums falling in the interval $(n, n + 10)$ was increasing with n. And, of course, much stronger partial results on the Goldbach conjecture have been obtained. Again, for the closely related twin prime conjecture (one of those conjectures that could not possibly be refuted by inspecting samples of numbers), rather compelling heuristic considerations[19] lead to the estimate

$$Z(N) \sim 2\pi \left(1 - \frac{1}{(1-p^2)}\right) \int_2^N (\ln n)^{-2} \, dn$$

for the number $Z(N)$ of twin primes $\leqslant N$. (Here the infinite product is taken over all odd primes p and converges to 0.66016 . . . .) Since the integral tends to infinity with N, the twin prime conjecture is 'proved'. The estimate of $Z(N)$ has been tested empirically and found to be very accurate up to 40 million. E.g., $Z(37 \times 10^6) = 183728$, while the approximation gives 183582.

Now if Popper's point is that no examination of 'positive cases' could ever raise the probability of such a conjecture to a finite positive value, I cannot but agree. Instances alone can never sway us! But if his claim is that *no evidence of any kind* (short of actual proof) can raise the probability of a general law to a finite positive value, I emphatically disagree. On the basis of the cited evidence for the twin prime conjecture, for example, it would seem to me quite rational to accept a bet on the truth of the conjecture at odds of, say, 100: 1, that is, to stake say $100 against a return of $10 000 should the conjecture prove true.

I have focused on arithmetical conjectures because, it seems to me, we are far clearer about what we mean by calling them 'true' than we are about empirical laws or theories. Moreover, we know, on logical grounds alone, that every even number is a sum of two, or three, . . . odd primes (where our convention to insure logical exclusiveness is assumed). In empirical science, on the other hand, we never really know that a partition of hypotheses exhausts the possibilities. This dependence is easily lost sight of, but it is crucial for interpreting both the likelihood principle and prior probabilities. The latter must be construed (in empirical contexts) as *relative* weights, or taken conditional on the assumption of exhaustiveness. Taken in that spirit, they are quite useful for comparing hypotheses. Where hypotheses are not exclusive, or where the likelihood function cannot be computed, the method to be developed in Chapter 9 can be employed.

NOTES

Abbreviations: *LSD* for Popper (1959), C&R for Popper (1963) and Schilpp, for Schilpp (1974).

[1] *C&R*, p. 390.
[2] Compare Popper's comment in *Schilpp*, p. 1192, Note 165b.
[3] Jeffrey, (1975), pp. 111–112.
[4] *LSD*, pp. 401–2, 418: Popper argues that, like the requirement of total evidence, that of sincerity cannot be formalized.
[5] There are well known cases in psychology where the experimenter's theoretical expectations influence the behavior of his subjects (cf. R. Rosenthal, *Experimental Effects in Behavioral Research*, Appleton-Century-Crofts, 1966), but these exceptions prove the rule, for it is the experimental outcomes themselves that are affected, not just their interpretation.

[6] Good (1975a).

[7] Confusion on this point has arisen due to Popper's having published a proposed quantitative measure of degree of corroboration. The purpose of the measure, however, was to show that corroboration is not a probability (or even a probability increment, in view of the 'paradox of ideal evidence') and to establish the consistency of a set of proposed desiderata (*LSD*, appendix *ix, esp. p. 387). But I think the real bone of contention here is not over the virtues of formalization *per se* but over Popper's contention that *not all* essential features of corroboration can be formalized (cf. Note 4 above). The features Popper cites all involve the experimenter's intentions, etc., and since the likelihood function is uninfluenced by the experimenter's state of mind, Bayesians deem these features irrelevant.

[8] Cf., e.g., *Schilpp*, 990–993, for a very full statement of the argument.

[9] *C&R*, pp. 247–8. Awkward consequences of predesignationism in statistics are set out in Chapters 10, 11.

[10] If the observed angle of deflection of light rays in the gravitational field of the sun had fallen outside an allowed margin of error laid down in advance (though only slightly so), that would certainly have called for correction of, but not wholesale rejection of, General Relativity. That is, we would be loath to reject a theory that predicted an effect so surprising as that, if only with imperfect accuracy. This would be a good example of an 'inaccurate but improbably accurate' theory (see the end of Section 5 below and Note 2 above).

[11] *Schilpp*, p. 1193, Note 166.

[12] *LSD*, Section 22, esp. p. 87.

[13] This aspect of physical theories has been emphasized in conversation by J. Sneed, as well as in a recent talk delivered at Stanford ('Is Quantum Mechanics Applied Probability?'). Cf. also Sneed (1971).

[14] *C&R*, p. 241, Note 24, where Popper proposes relativising simplicity to a problem; his notion of 'dimensionality', of course, is relativised to a 'field of relatively atomic statements' (*LSD*, appendix *viii, esp. p. 379). I take my 'adaptation', then, to be in something of a Popperian vein.

[15] *LSD*, Section 20 and *C&R*, Chapter 10, Section 5.

[16] For more on this, see Chapter 5.

[17] Arguments to the effect that general laws have probability zero in infinite domains given *finite* samples are marshalled in *LSD*, appendix *vii. Suitably qualified, I do not disagree with this conclusion (see below). But Popper seems to be hankering after the much stronger conclusion that no evidence of any kind (short of proof) could raise the probability of a general law above zero, and his arguments certainly won't support that much.

[18] If you assign zero probability to logically possible outcomes, you can be made the victim of a quasi Dutch Book, viz., an arrangement of stakes such that, betting at the corresponding odds, you cannot do better than break even whatever outcome occurs.

[19] Shanks (1962), pp. 30, 214 (ex. 379). Hardy and Wright (1960) give a similar estimate of $Z(N)$ – Section 22.20 – from which the infinitude of twin primes also follows. Both works contain many passages in which the authors express high degrees of credence or assign high probability to various unproved conjectures.

BIBLIOGRAPHY

Ackermann, R.: 1976, *The Philosophy of Karl Popper*, University of Massachusetts Press, Amherst, Massachusetts.

Good, I. J.: 1967, 'The White Shoe Is a Red Herring', *Brit. J. Phil. Sci.* **17**, 322.

Good, I. J.: 1975, 'Explicativity, Corroboration, and the Relative Odds of Hypotheses', *Methodologies: Bayesian and Popperian, Synthese* **30**, 39–73.

Hardy, G. H. and E. M. Wright: 1960, *An Introduction to the Theory of Numbers*, fourth edition, Oxford University Press, Oxford.

Hintikka, K. J. J.: 1967, 'Introduction by Elimination and Induction by Enumeration', in *The Problem of Inductive Logic* (ed. by I. Lakatos), Amsterdam.

Jeffrey, R. C.: 1975, 'Probability and Falsification: Critique of the Popperian Program', *Methodologies: Bayesian and Popperian, Synthese* **30**, 95–117.

Kuhn, T.: 1957, *The Copernican Revolution*, Cambridge, Massachusetts.

Kuhn, T.: 1962, *The Structure of Scientific Revolutions*, Chicago.

Lakatos, I.: 1970, 'Falsification and the Methodology of Scientific Research Programmes', in *Criticism and the Growth of Knowledge* (ed. by I. Lakatos and A. Musgrave), Cambridge.

Popper, K.: 1959, *The Logic of Scientific Discovery*, London.

Popper, K.: 1963, *Conjectures and Refutations*, London and New York, Basic Books, New York.

Provine, W. B.: 1971, *The Origins of Theoretical Population Genetics*, Chicago.

Rosenkrantz, R. D., 1973, 'Probabilistic Confirmation Theory and the Goodman Paradox', *Amer. Phil. Quart.* **10** (1973), 157–162.

Rosenkrantz, R. D.: 1976, 'Simplicity', *Foundations of Probability and Statistics and Statistical Theories of Science* (ed. by C. K. Hooker and W. Harper), Vol. 1, D. Reidel, Dordrecht, Holland, 167–203.

Schilpp, P. A. (ed.): 1974, *The Philosophy of Karl Popper*, Open Court Publishing Co., LaSalle, Illinois.

Shanks, D: 1962, *Solved and Unsolved Problems in Number Theory*, Vol. 1, Spartan Books, Washington, D.C.

Sneed, J.: 1971, *The Logical Structure of Mathematical Physics*, D. Reidel, Dordrecht, Holland.

CHAPTER 7

THE COPERNICAN REVELATION

1. INTRODUCTION

"Judged purely on practical grounds", Thomas Kuhn has written,[1] "Copernicus' new planetary theory was a failure; it was neither more accurate nor significantly simpler than its Ptolemaic predecessors". Although Kuhn undertakes no detailed comparison of either theory with actual planetary positions, his claim about accuracy seems essentially correct (see the appendix). Nor can it be doubted that the theory of Copernicus was complicated. Some of the complications were forced on him by bad data; no reasonably simple theory could have fitted both the ancient and more recent observations Copernicus had at his disposal. On the other hand, at least some of the complications were unnecessary and can be charged to Copernicus himself. Those who maintain that the simplicity of the Copernican theory is a myth — what Robert Palter has aptly dubbed the '80–34 myth', referring to the number of circles each theory supposedly requires — have so much truth on their side. Neither could the 'novel' predictions of the new theory, stellar parallax and the motion of the earth, be directly confirmed. If Galileo set out to demonstrate the motion of the earth in his *Dialogue*, his attempt must be judged a failure. (True, he was able to detect the phases of Venus, but that observation is compatible with the Tychonic version of geocentrism.) Indeed, it is frequently urged that, from a relativistic standpoint, it makes no difference whether we refer the motions of the planets to the earth or the sun. Not only were objective grounds for preferring the heliostatic theory lacking at the time, no such grounds exist — or so it is claimed.

We are led by such considerations to a curious pass. It appears we must regard it as something of a happy accident that men to whom the new astronomy had an *aesthetic* appeal were good enough *propagandists* to establish it as worthy of serious consideration, thus paving the way to the Newtonian synthesis. The time has come to reassess this view, and with it the contribution of Copernicus himself.

2. COPERNICUS'S ARGUMENT

It has been surmised[2] that Copernicus was led to the heliostatic hypothesis
by noticing that the periods of the deferents of the inner planets and the
epicycles of the outer planets in Ptolemy's system are all equal to one year,
while the periods of the epicycles of the inner planets and the deferents of
the outer planets are equal to their synodic periods (the time between
successive oppositions to the sun). These relations clearly suggest that for an
outer planet the deferent acts as a planetary orbit around the sun, while for
an inner planet the roles are reversed. The situation for an inner planet is
depicted in Figures 1–3.

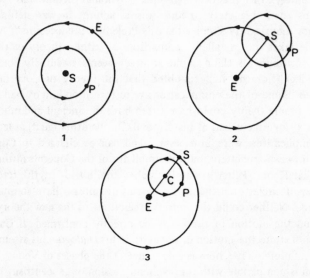

Figs. 1–3.

In the Ptolemaic theory (Figure 3), the epicyclic center C is constrained to
lie on the line ES joining the center of the earth to the center of the sun at all
times. If we make the additional assumption that S and C coincide (so that the
epicycle is sun-centered), we obtain the Tychonic system of Figure 2 as a
special case. This system can be transformed into an equivalent heliostatic
picture by the simple device of fixing S and sending E into orbit around S
(Figure 1), an orbit that is the reflection of S around E in Figure 2. The
period of the deferents in Figures 2 or 3 is seen to be that of the earth's

revolution around the sun. I shall not treat the case of an outer planet in detail, as exactly similar considerations apply.

Now let us reverse the steps that took us from Figure 3 to Figure 1. By fixing E in Figure 1, we obtain the equivalent Tychonic system of Figure 2, which is, in turn, a special case of the general Ptolemaic theory of Figure 3. Since the heliostatic picture of Figure 1 and the Tychonic picture of Figure 2 are equivalent (so far as the planets are concerned), we see that Figure 1 can be regarded as a special case of Figure 3. All the special cases of Figure 3 agree in their predicted angular variations of the planets but differ in the location of the epicyclic center C on the line ES (for the inner planets). As we saw in Chapter 5, because the average of the likelihood function can never exceed its maximum, a theory can never be better supported than its best-fitting special case. Hence, the heliostatic theory of Figure 1 (be it based on uniform circles, eccentric circles, or Kepler ellipses) is at least as well supported as the corresponding geocentric theory of which it is a special case – taking only angular variations into account. For the observations initially could not determine the location of C on ES, and so all special cases of Figure 3 fit the angular variations equally well. The likelihood function is thus a constant, and hence, the heliostatic theory is exactly as well supported by the angular variations as the geostatic theory, but no more so.

Matters are quite otherwise when variations in apparent sizes of the planets are taken into account.[3] When it is on the far side of the sun (superior conjunction), Venus is slightly more than six times more distant from the earth than it is at inferior conjunction, and hence its apparent size (i.e., surface area) should be nearly forty times as great at inferior conjunction. But the naked eye reveals not even a doubling of its apparent size. This, and not the undetectability of stellar parallax, seems to me to have been the strongest purely astronomical objection to Copernicus's innovation. One might think it so strong, indeed, that Copernicus should have given up his hypothesis when confronted with it. But he had other weighty reasons for thinking his hypothesis correct (Section 3), and his steadfastness in the face of Venus's recalcitrance provoked Galileo to the following paean:

Nor can I ever sufficiently admire the outstanding acumen of those who have taken hold of the opinion [that the earth moves] and accepted it as true; they have through sheer force of intellect done such violence to their own senses as to prefer what reason told them over that which sensible experience plainly showed them to the contrary. For the arguments against the whirling of the earth which we have already examined are very plausible, as we have seen; and the fact that the Ptolemaics and Aristotelians and all their disciples took them to be conclusive is indeed a strong argument of their effectiveness. But the experiences which overtly contradict the annual movement are indeed so much

greater in their apparent force that, I repeat, there is no limit to my astonishment when I
reflect that Aristarchus and Copernicus were able to make reason so much conquer sense
that, in defiance of the latter, the former became mistress of their belief [*Dialogue*,
p. 328].

And later he adds:

For as I said before, we may see that with reason as his guide he resolutely continued to
affirm what sensible experience seemed to contradict. I cannot get over my amazement
that he was constantly willing to persist in saying that Venus might go around the sun
and be more than six times as far from us at one time as at another, and still look always
equal, when it should have appeared forty times larger [*Dialogue*, p. 339].

To actually calculate the relevant likelihoods would require an assumption
about the reliability with which apparent sizes (seen by the naked eye) reflect
astronomical distances. Since the sun and moon appear much larger near the
horizon, such observations are clearly not altogether reliable, and this may
have reconciled Copernicus to the anomalous lack of increase in the apparent
size of Venus. At any rate, as Galileo says, he had other powerful reasons for
preferring the heliostatic model.

The situation changed dramatically with the invention of the telescope. At
inferior conjunction, Venus is horned and palely lit by the sun, while at
superior conjunction, it is directly lit. To the naked eye, it consequently
appears to be of roughly equal size at its two conjunctions. Galileo first
observed that, seen through a telescope, the halo of diffused light is stripped
away, and then Venus does indeed appear about forty times larger at inferior
conjunction. At this point, the heliostatic (or kinematically equivalent
Tychonic) theory becomes *the* best-fitting special case of the geostatic
theory, given both the angular variations and variations in apparent sizes. I
think Galileo regarded this, together with the phases of Venus which the
telescope also revealed, as the clinching argument for Copernicanism. (In
actual fact, of course, it only narrows the field to the Copernican and
Tychonic alternatives.) "O Nicholaus Copernicus", he writes (*op. cit.*,
p. 339), "what a pleasure it would have been for you to see this part of your
system confirmed by so clear an experiment!" Galileo was right to make
much of this point (even though he ignored the Tychonic alternative). As the
linkage example of Chapter 5 illustrates, when large samples or precise
quantitative predictions are in question, the average likelihood of a
best-fitting special case will typically exceed that of the general theory by an
enormous amount.

But how was the prediction in question obtained? On the face of it, the heliostatic model requires essentially no more than that the epicyclic center C coincide with the center of the sun, as in Figure 2. Nothing is said, however, about the absolute sizes of the circles. This apparent freedom of the parameters disappears, though, when account is taken of the observations. Figure 4 (from Kuhn (1957)) readily suggests the elementary geometric argument by which the radius of Venus's orbit is determined in astronomical units (where 1 astronomical unit is the earth's mean distance from the sun). At maximal elongation, E, V, and S form a right triangle, and the angle at E can be directly measured. It requires only a slightly more elaborate argument to determine the radii of orbits of outer planets in astronomical units (Kuhn, 1957, p. 176). Similar determinations can be made, moreover, for non-circular orbits, and are made routinely by both Ptolemy and Copernicus.

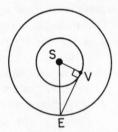

Fig. 4.

No like determination of relative distances is forthcoming in Ptolemaic astronomy, unless, of course, we make the assumption, unwarranted from a geostatic point of view, that C and S coincide. For, omitting S from Figure 3, the whole picture can be subjected to an arbitrary magnification or contraction (i.e., a similarity transformation) without affecting angular variations. Hence, the observations determine only the ratio of the distances EC and CV in Figure 3, and not their lengths in astronomical units (unless $C = S$).

The upshot, then, is that the geostatic theory has *two* additional degrees of freedom when the observations are taken into account — even if we assume, with Ptolemy, that C must lie on the line ES. First, we can choose at will the location of C on ES, and secondly, we can freely adjust the absolute sizes of the circles employed.[4] Copernicus quite properly stresses this difference between the theories in his preface:

Thus, assuming motions, which in my work I ascribe to the earth, by long and frequent observations I have at last discovered that, if the motions of the rest of the planets be brought into relation with the circulation of the earth and be reckoned in proportion to the circles of each planet, not only do their phenomena presently ensue, but the orders and magnitudes of all stars and spheres, nay the heavens themselves, become so bound together that nothing in any part thereof could be moved from its place without producing confusion in all the other parts of the universe as a whole.

The geometric argument of Figure 4 also shows that Mercury is closer to the sun than Venus, long an unsettled point of controversy in the geostatic scheme. Ptolemaic astronomers were forced to fall back on some arbitrary assumption or other, and the most commonly favored was the rule of thumb that a planet's distance from the center is proportional to its period. That assumption fails for the inner planets and the sun, since all their deferents have the same period (a year).

These simplifications of the Copernican system (by virtue of which it is a *system*) are frequently cited reasons for preferring it to the Ptolemaic theory. Yet, writers from the time of Copernicus to our own, have uniformly failed to analyse these simplifications or account adequately for their force. The Bayesian analysis of Chapter 5 fills this lacuna in earlier accounts by showing that the cited simplifications render the heliostatic theory (or the equivalent Tychonic theory) *better supported* than the corresponding (more complicated) geostatic theory. By allowing fewer possible planetary orbits, the heliostatic theory is better supported by its agreement with the orbits observation discloses.

This consideration equally supports the Tychonic system, as I have admitted all along. But while the heliostatic system of Figure 1 and the Tychonic system of Figure 2 are equivalent, qua theory of the planets, they are inequivalent theories of the heavens. The Tychonic theory permits retention of a central, motionless earth (its *raison d'être*), and thereby averts the strong objection to the Copernican hypothesis based on the unobservability of stellar parallax and the undetectability of the earth's motion. Tycho saw that the actual sizes of the stars, given their apparent sizes and the distance they must be for their parallaxes to go undetected, must be enormous. Even a star of the third magnitude was assumed to have a diameter of $65''$ of arc and would therefore have to be comparable to the earth's assumed orbit around the sun in size. By reducing stars to mere points of light, the telescope erased this objection too, but that happened more than thirty years after Tycho proposed his theory and his objection.

On the other hand, the supposed Achilles heel of the Copernican theory — the enormous lower bound it imposes on the distance to the nearest star — may have been the one reason of substance for preferring it to the Tychonic theory, which imposes no constraint on this parameter (beyond that already imposed by triangulation measurements). Tycho estimated that a moving earth required the distance from Saturn to the stars to be about seven hundred times the distance of Saturn from the sun. He did not consider this distance implausibly large because he took seriously the time-honored but tenuous estimates of the distance based on an assumption of closest packing of the celestial spheres, but because that distance, as we have seen, would require the stars to be implausibly large if Copernicus' hypothesis were correct.

The situation that confronted astronomers in the transitional period from 1543 to 1609, then, was, roughly speaking, that the additional constraint which the Copernican theory imposed on the distance to the stars was a rather implausible constraint. Given, too, the apparent force of the physical arguments against the earth's motion, the evidence would appear to have favored the Tychonic theory. And, of course, it was a very widely favored theory for a time. Although there were a number of convinced Copernicans during this period — Rheticus, Bruno, Recorde, Digges, Rothmann (who defended the theory ably against the objections of Tycho), Gilbert (who refers in *De Magnete* to the 'daily magnetic revolution of the earth'), and Maestlin among them — the fact remains that truly decisive evidence for the heliostatic theory was not available until Bradley detected the aberration of light, Bessel detected stellar parallax, and Foucault designed his pendulum. One should not conclude, nevertheless, that the considerations which prompted acceptance of the Copernican scheme long before these pieces of evidence were available were necessarily extra-evidential.

By undermining the physical arguments aimed at showing the earth does not (or cannot) move, Galileo increased the plausibility of the Copernican theory. Similarly, his telescopic observations undercut Tycho's argument that the stars must be implausibly large for their annual parallaxes to go undetected, and thereby rendered less implausible the additional constraint Copernican theory imposes on the distance to the stars. Indeed, because planets appear as flat discs and stars as mere points of light through a telescope, a very much larger spatial gap is indicated between stars and planets. Even Ptolemaic estimates of the distance were immense by ordinary terrestrial standards and Ptolemy himself describes the earth as a mere point as compared with the stellar sphere. Given, on the other hand, even the minimal distance to the stars suggested by triangulation measurements, the linear velocity of a diurnally rotating star would be such that the star could circuit the earth's surface in a few seconds. Given the actual distance, the linear velocity of a diurnally rotating star would exceed that of light *in vacuo*, and so the argument which runs, "from a relativistic standpoint, it doesn't matter whether the earth is assumed to be in motion or not", does not apply, at any rate, to the diurnal rotation of the earth. Copernicus and his followers made much of this point: that the immense distance to the stars their theory required was no more implausible than the immense speed of diurnally rotating stars required by Ptolemaic theory. I shall have more to say about reasons for preferring the Copernican system to the Tychonic in Section 4 below.

3. MANY EFFECTS; ONE CAUSE

It would be well to open this section with a passage from Lakatos and Zahar (1976):

Having specified that the unit of a mature science is a research program, I now lay down the following rules for appraising programs. A research program is either progressive or degenerating. It is *theoretically progressive* if each modification leads to new unexpected predictions and it is *empirically progressive* if at least some of these novel predictions are corroborated. It is always easy for a scientist to deal with a *given* anomaly by making suitable adjustments to his program (e.g., by adding a new epicycle). Such manoeuvres are *ad hoc*, and the program is *degenerating* unless they not only explain the given facts they were intended to explain but also predict some new fact as well.

Elsewhere, Lakatos tells us that a prediction is 'novel' if it goes against prevailing background knowledge, or better, if it is excluded by a rival research program. By enlarging sample coverage and reducing testability, complications of a theory can rarely be said to issue in 'novel' predictions. Yet, a complication of a theory may improve its accuracy sufficiently to increase its average likelihood and so constitute 'progress' from a Bayesian point of view.

Lakatos is constrained to recognize that many theories are strongly 'corroborated' by facts or low level empirical laws known long before the theory was formulated. One has only to mention the Balmer formula and Periodic Law, the laws of constant and multiple proportions, or the advance of the perihelion of Mercury. To admit 'corroboration' of a theory by a known law, Lakatos falls back on a suggestion of Elie Zahar's which counts predictions as 'novel' when the theory in question was not expressly formulated to account for them (when it explains them in passing, as it were).

Now this manoeuvre has all the earmarks of a 'conventionalist stratagem' or 'degenerating problem shift', but I needn't press this point, for, even in the refined sense, I fail to see what 'novel' predictions flow from, say, the replacement of uniform sun-centered circles by eccentrics uniform about an equant, or the replacement of Mendel's original laws by polygenic models with complexly interacting systems of genes. Why should it matter whether or not Bohr expressly formulated his theory of the atom to explain the Balmer formula (rather than, as Lakatos somewhat dubiously claims, to explain the stability of the Rutherford atom)? Why should it make a difference whether or not Hipparchus introduced epicyclic motions to explain the variations in the brightness of the planets rather than merely to fit their observed retrogressions? It is hard to see why psychological facts about what motivated a theory should affect its status.

My difficulties with the refined sense of 'novel' are multiplied when Lakatos comes to apply this notion to the many irregularities of planetary motion that are immediate consequences of the earth's motion. The stations and retrogressions of the planets, the phases of Venus, its variation in apparent brightness and diameter, the bounded elongation of an inner planet, and so on, all count now as novel predictions of the Copernican theory but not of the Ptolemaic. Lakatos observes that "although these facts were previously known, they lend much more support to Copernicus than to Ptolemy within whose system they were dealt with only in an *ad hoc* manner, by parameter adjustment" (*op. cit.*, p. 376). That is to say, the Ptolemaic theory can account for these effects only by fitting additional parameters or by making additional assumptions. To thus characterize the difference, however, is pretty clearly to give the game away. We see now that it has nothing whatever to do with the suspect notion of 'novelty' and everything to do with the (wholly objective) simplicity or overdetermination of the heliostatic theory. "We thus rather follow Nature", Copernicus writes, "who producing nothing vain or superfluous often prefers to endow one cause with many effects". Farther on (*De Rev.*, Bk. I, ch. 10) he continues:

For here we may observe why the progression and retrogression appear greater for Jupiter than Saturn, and less than for Mars, but again greater for Venus than for Mercury; and why such oscillation appears more frequently in Saturn than in Jupiter, but less frequently in Mars and Venus than in Mercury; moreover, why Saturn, Jupiter and Mars are nearer to the earth at opposition to the sun than when they are lost in or emerge from the sun's rays. Particularly Mars, when he shines all night appears to rival Jupiter in magnitude, being distinguishable only by his ruddy color; otherwise he is scarce equal to a star of the second magnitude, and can be recognized only when his movements are carefully followed. All these phenomena proceed from the same cause, namely Earth's motion.

Let us provisionally label this virtue of the theory *economy* or *economy of explanation*. Some would regard economy as an 'epistemic utility' to be weighed apart from a theory's agreement with the data or its evidential support. Others might hold that it is a merely aesthetic desideratum, and, as such, no compelling ground for preferring one theory to another. Let me quote what appears to be an especially strong statement of the latter position from Kuhn (1957), p. 181:

It is through arguments like these that Copernicus seeks to persuade his contemporaries of the validity of his new approach. Each argument cites an aspect of the appearances that can be explained by *either* the Ptolemaic *or* the Copernican system, and each then proceeds to point out how much more harmonious, coherent, and natural the Copernican explanation is. There are a great many such arguments. The sum of the evidence drawn from harmony is nothing if not impressive.

But it may well be nothing. 'Harmony' seems a strange basis on which to argue for the earth's motion, particularly since the harmony is so obscured by the complex multitude of circles that make up the full Copernican system. Copernicus's arguments are not pragmatic. They appeal, if at all, not to the utilitarian sense of the practicing astronomer but to this aesthetic sense and to that alone. They had no appeal to laymen, who, even when they understood the arguments, were unwilling to substitute minor celestial harmonies for major terrestrial discord. They did not necessarily appeal to astronomers, for the harmonies to which Copernicus's arguments pointed did not enable the astronomer to perform his job better. New harmonies did not increase accuracy or simplicity. Therefore they could and did appeal primarily to that limited and perhaps irrational subgroup of mathematical astronomers whose Neoplatonic ear for mathematical harmonies could not be obstructed by page after page of complex mathematics leading finally to numerical predictions scarcely better than those which they had known before.

I have no wish to deny the powerful aesthetic appeal of the cited 'harmonies'. I do insist, however, that aesthetic responses ought to sway us only when they point to some deeper lying simplification or feature which renders a theory more confirmable in the light of conforming data. What I am calling the 'economy', what Lakatos calls the 'novelty', and what Kuhn calls the 'harmony' of the Copernican system is nothing but an aspect of its simplicity or overdetermination. Each of the consequences of the earth's motion which Copernicus exhibits is an aspect of planetary motion which the heliostatic theory *determines*, but which can be accommodated within a geostatic framework only by fitting adjustable parameters or by making additional assumptions.

Consider first the decreasing frequency of retrogression in the sequence Saturn, Jupiter, Mars. Saturn completes one trip through the zodiac in not quite 30 years, Jupiter in not quite 12, and Mars in not quite 2 (and these must be their sidereal periods as well). Hence, the earth overtakes Saturn nearly 30 times, Jupiter nearly 12, and Mars nearly twice in one trip through the zodiac, so that, as seen from Earth, the orbit of Saturn around the earth has almost 30 loops, Jupiter's has almost 12, and Mars's has almost 2. In terms of the heliostatic model, then, it is clear why these frequencies of retrogression characterize the Ptolemaic orbits. On the other hand, within the Ptolemaic theory, the period of an outer planet's epicycle can be adjusted at will to produce as many retrogressions in one circuit of the zodiac as desired. Given the observed time of a trip through the zodiac (the sidereal period and the period of the Ptolemaic deferent), the frequency of retrogresssion (or the number of stations) is determined heliostatically but not geostatically. Hence, while both theories fit this aspect of the data, the Copernican theory fits *only* the actually observed frequency of retrogression for each planet. Otherwise put, the heliostatic model determines the period of the Ptolemaic epicycle as a function of the epicycle's deferent, but within the geostatic scheme, these periods are independently adjustable. Again, that makes the heliostatic (or kinematically equivalent Tychonic) theory a special case of the corresponding geostatic theory for the pertinent aspect of the data. Hence, while both theories fit the data equally well, qua special case of the geostatic theory, the heliostatic theory is again better supported.

Exactly the same analysis shows that the heliostatic theory is better supported by each of the other aspects of planetary motion which Copernicus cites. The order of even the outer planets can only be set in Ptolemaic theory by applying the (essentially arbitrary) rule of thumb that the larger circle has the longer period. Even if we fix the order as Ptolemy does, there is no reason to expect the angular length of the retrogression of Mars to exceed that of Jupiter (i.e., no reason why the Ptolemaic epicycle of Mars should be so large compared with that of Jupiter or Saturn). But this does follow heliostatically, for the parallax of a nearer object is larger. (This, in turn, determines that Mars is closer to the earth than Jupiter.) Again, Mars has greater angular velocity than Jupiter (as determined by their sidereal periods), so that the earth will overhaul it less rapidly, thereby also explaining the longer time Mars spends retrogressing. Nor could the Ptolemaic system account for the strange coincidence that the three outer planets should all be at the perigee of their epicycles when at opposition to the sun. That each of these appearances follows by 'mere mathematical necessity' from the assumption of a moving earth is advanced by Kepler too (in the opening chapter of his *Mysterium Cosmigraphicum*) as the decisive reason for preferring the Copernican theory to the Ptolemaic. Indeed, it is the 'other reason' to which Galileo alludes in the passage quoted above, and as Dreyer says (Dreyer, 1953, p. 373), "it is difficult to see how anyone could read this chapter [of Kepler's] and still remain an adherent of the Ptolemaic system". Dreyer then goes on to say:

What Kepler aimed at throughout his whole life was to find a law binding the members of the solar system together, as regards the distribution of their orbits through space and their motions, knowing which law he expected it would be possible to compute all the particulars about any planet if the elements of one orbit were known.

Kepler makes it abundantly clear that the simplification Copernicus effected lies, not in the reduction of circles *per se*, but in the coherent explanations of the major irregularities of planetary motion that are forthcoming in his theory. It is the fact that the Copernican model is able to accommodate *only* the actual observations that renders its explanations genuine and shows that the irregularities themselves are *mere* appearances or effects of the earth's motion. Kepler would not have agreed, then, with Kuhn's contention that either theory could explain the irregularities of planetary motion; he would only have admitted that either theory could be made to fit these irregularities.[5]

So much for novelty. But Lakatos lays down another requirement of 'heuristic progress': that the successive modifications of what he calls the 'protective belt' of hypotheses surrounding the core assumptions of a theory be "in the spirit of the heuristic" (p. 369) and he adds (p. 371) "the use of the equant was tantamount to abandonment of the Platonic heuristic". Now there can be no doubt that Copernicus attached great importance to the elimination of equants, for he believed that uniform circular motion is the motion natural to a sphere.[6] In practice, though, he merely replaced the Ptolemaic eccentric-equant construction with an equivalent epicycle construction (as shown in Figures 6–7 of the appendix). Moreover, he was perfectly clear about their equivalence and recognized as open the question which of the two mechanisms nature actually employs. What interests me in this connection is that Kepler strongly preferred the eccentric-equant construction of Ptolemy, and *just because it had greater heuristic value*. He found it increasingly difficult to physically interpret epicyclic motions. Sensing the need for a more adequate dynamics, he assumed the planets are driven in their orbits by a force emanating from the sun. It was also natural to assume that this force diminishes (linearly) with distance, and that would account for the empirical rule (true without exception in a heliocentric system) that planets farther from the sun have longer orbital periods. Kepler reasoned that this diminution of the force with distance from the sun should also govern a single planet in its orbit around the sun. In that case, there is a *real non-uniformity* in the planet's motion, a non-uniformity that is brought out by the eccentric-equant picture but obscured by the epicycle picture which tends to reduce it to the status of an appearance resulting from a combination of uniform circular motions.[7]

I suppose, if it came to that, one would have to reckon Newton's use of 'fluxions', Einstein's use of Riemannian geometry, and the introduction of quanta as departures from (or not 'in the spirit' of) a prevailing heuristic. But then it is just *mysterious* why Lakatos's vague requirement of heuristic progress should count for anything (just as it is mysterious why 'novel' predictions, in the extended Zaharian sense, should count for anything). One is left wondering whether this methodologizing from on high can advance our understanding of science any more than the mythologizing of science.

Similar remarks apply to 'epistemic utilities'. It is one thing to single out given desiderata or 'utilities', another to show why they ought to be taken seriously, and still another to show how they are to be weighted in an overall appraisal of a theory. On the other hand, we have seen that the 'economy' of the heliostatic scheme is but an aspect of its simplicity or overdetermination, and that the latter tends to be reflected in higher evidential support. That the simpler of two equally accurate theories is better supported, while not an invariable rule, is true in a broad enough spectrum of cases to shed real light on rational theory appraisal and rational theory change. It is not clear, by contrast, whether the approach through either epistemic utilities or imposed methodological rules amounts to more than a convenient labelling of those features of theories we sense are important without knowing why.

I would urge, in stark contrast to these viewpoints, that rational theory appraisal is governed solely by the evidence at hand and the comparative support it accords the competing theories. 'Epistemic utilities', like simplicity, enter only indirectly as contributing to the support that accrues to an accurate theory. Certainly nothing I have said about the Copernican Revolution even hints that extra-evidential considerations largely, or even partially, shaped the beliefs of professional astronomers at any given time. I am suggesting, on the contrary, that the beliefs of scientists tend to mirror the evidence. In the next and final section, I will examine the evidence from terrestrial physics and attempt to explain why even the Tychonic theory attracted few adherents after 1632, the year Galileo published his *Dialogue*. My account, up to this point, has at least the virtue of providing a clear, objective and not merely aesthetic basis for the very arguments Copernicus and his followers found most compelling. Against this, alternative articulations of those arguments in terms of vaguely specified epistemic utilities or arbitrary methodological rules seem strained, overly elaborate, and not really to the point.

4. THE PHYSICAL ARGUMENTS

Given the narrower choice between the Tychonic and Copernican alternatives, it is alleged that physical considerations militated strongly against the earth's motion and in favor of the Tychonic theory. It is commonly held, indeed, that the earth's motion was excluded by a well-entrenched physical theory — Aristotle's — and for this reason alone should have been assigned a low prior probability. The situation, however, was more complicated. Aristotle's theory of violent (or projectile) motion was challenged earlier on by the commentator John Philoponus (5th century A.D.), who offered an alternative account in terms of impetus. That account was revived and further articulated (in different versions) by John Buridan and Nicole Oresme in the fourteenth century, and Kuhn has written that "it replaced Aristotelian dynamics in the work of the principal medieval scientists" (Kuhn, 1957, p. 120). As Oresme noted (cf. the selection from Oresme in Chapter 67 of Grant (1974)), the physical *possibility* of the earth's motion could be made out on the basis of impetus theory, for the earth would impart its impetus to objects on or near its surface, and hence, contra Ptolemy and Aristotle, an arrow shot vertically upwards would return to the same spot on the moving earth. (Oresme makes much of the relativity of motion, and many of his arguments reappear almost verbatim in Bk. I of *De Revolutionibus*.)

Yet, strangely enough, Aristotelian arguments against the whirling of the earth continued to be employed throughout the transitional period I am considering, most notably by Tycho himself. The claim that a 'paradigm shift' had occurred becomes correspondingly difficult to swallow. Shapere (1974) views the impetus theory, instead, as "only one of many deviant movements within the broad framework of Aristotelianism" (p. 54). Wherever the truth may lie, it is clear that the impetus theory, while it offered an effective base from which to criticize Aristotle's treatment of 'violent motion', had obscurities and difficulties of its own.

The most critical difficulty concerns the rate at which impetus is dissipated. Buridan asserts that denser objects are capable of receiving more impetus. Shapere writes (*op. cit.*, p. 51): "Buridan says only that the quantity of matter determines how much impetus *can* be imparted to the body, not that it determines how much *is* imparted". On this interpretation, there is no telling how much impetus a body of given weight receives from a given motion, and, in particular, there is no telling whether the heavier of two bodies of the same size and shape and initial speed will travel farther. But Shapere's reading is a dubious one, for in almost the very next sentence from Buridan (translated in Clagett (1959), pp. 532–538 and reprinted in Grant (1974), Chapter 48), Buridan does say: "And so also if light wood and heavy iron of the same volume and the same shape are moved equally fast by a projector, the iron will be moved farther because there is impressed in it a more intense impetus . . .". To be sure, one could escape the conclusion that the heavier object travels farther while retaining the view that it receives more impetus from a given motion by saying that the greater gravity of such a body overcomes (or 'corrupts') its impetus more rapidly. Still, no serious attempt was made by Buridan or his followers to state a precise mathematical relation that would enable one to compute the range of a projectile as a function of what appear to be the relevant parameters: initial velocity, weight, size, shape, and the density of the medium. In retrospect, one recognizes the necessity of abstracting away some of these obfuscating variables in order to arrive at such a law.

According to Buridan, a body's gravity continually impresses additional impetus upon it, accounting for its acceleration. However, since the impetus a body receives from a given contact force is, like its weight, proportional to its 'quantity of matter', it seems to follow that heavier bodies not only fall faster but accelerate more. But neither effect is observed. Hence, while Buridan is able to account for acceleration in free fall, and perhaps even for the constancy of this acceleration (for each body), his theory leads to the unacceptable consequence that a wooden ball will have both a smaller acceleration and a smaller velocity in free fall than a lead or iron ball. With this much background, let us turn to some of the physical arguments which Tycho marshalled against the earth's motion.

Like Aristotle, Ptolemy and many others before him, Tycho argued that a stone dropped from a tower would not fall at the foot if the earth moved (just as, or so he thought, a stone dropped from the mast of a moving ship would not fall at the foot, but would be left behind), and further, that a cannon ball shot westward would travel farther than one shot eastward, since the cannon would be moving in the same direction as the cannon ball in the latter case but receding from it in the former.

Although the same principles are involved, the second argument was somewhat novel and seemed, superficially at least, to make as much trouble for the impetus theory as for Aristotle (given that the earth rotates). A point on the earth near the equator has a linear speed of the order of 1000 miles per hour, and so it is hard to see how anyone standing there could project a stone westward fast enough to overcome the contrary impetus imparted by the moving earth. But, in fact, an impetus theorist could resolve the difficulty by making the very natural assumption that impetus is additive. E.g., if a musket ball is fired from a moving coach, the additional impetus which the musket imparts to the ball is unaffected by the motion of the coach and would be the same whether the coach were (uniformly) in motion or at rest. It still doesn't follow, of course, that the ball will travel the same distance from the coach in either case. For that we need the additional assumption that the impetus the coach imparts carries the ball the same extra distance which the coach itself covers in the time of flight. But for this to be true of projectiles of different material and density virtually *requires* that the impetus a given motion imparts be independent of the density, gravity or 'quantity of matter' of the projectile. It would then follow that its density does not affect the rate at which a body falls or accelerates. Again, one could argue that the impetus impressed by gravity, like that impressed by a contact force, is proportional to a body's weight, so that bodies of the same size, shape and initial velocity, but differing in weight or density, would travel the same distance. Nevertheless, it is just as simple and squares better with the appearances to suppose that bodies containing different quantities of matter in the same volume *receive the same impetus from the same motion* and so fall at the same rate, and, moreover, that *the extra motion imparted by a given motion be equal to that given motion.* From this it is but a short step to the conclusion that in 'compound motions' generally, the components of motion are independent (and, in particular, the vertical and horizontal components). Moreover, the assumptions about compounded horizontal motions agree with what is observed in the special case where no additional motion is imparted. That is, if musket balls of different density (but the same size and shape) received a different impetus from the same motion, not all of them would remain stationary on the floor of a uniformly moving ship.

I don't claim for a moment that Galileo was led to his laws of free fall and projectile motion by such considerations, although he could have been. More likely, reflections on the relativity of motion, such as one finds in Oresme and Copernicus, led him there. Since there is no way of telling which of two ships out of sight of land and in uniform translational motion with respect to each other is 'really' in motion, a stone dropped from the mast should describe the

same (straight vertical path) with respect to each ship, for, otherwise, there
would be a way of determining which ship was 'really' in motion and which at
rest [cf. the *Dialogue*, pp. 186–187]. Seen from the shore, though, the stone
would describe a curved path, the resultant of its vertical and horizontal
motions, and it is easily shown that the curved path in question is a
semi-parabola, like that described by the water in a fountain.

Given the assumptions that the components of a compounded motion are
additive and independent, Tycho's east—west cannon shot falls harmlessly to
the ground. But Galileo went farther and turned the tables on Tycho here by
turning the cannon shot through a ninety degree angle. The resulting
north—south cannon shot is depicted in Figure 5a. A cannon ball is shot from

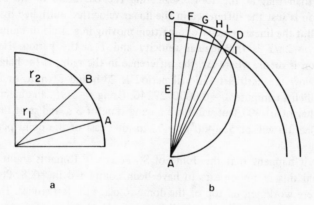

Fig. 5a–b.

B to a point A due south of B. At the more northerly latitude, the radius r_2
of the circle of latitude is smaller than the radius r_1 at A. Given the equality
of angular velocity at the two points, the target A moves farther eastward
(since it travels on a larger circle and has, therefore, larger *linear velocity*)
than the cannon at B in the time of flight. Hence, an apparent *westward*
deflection of the cannon ball will be observed (as Tycho maintained). Galileo
had this argument – his sought-for *experimentum crucis* – in hand when he
wrote that "the cannon will sometimes be placed closer to the pole than the
target and its motion will consequently be slower, being made along a smaller
circle", but then he adds "this difference is insensible because of the small
distance from the cannon to the mark" [*Dialogue*, p. 179]. Doubtless, it
would have been difficult to reliably detect the predicted deflection with the
artillery pieces available in Galileo's day, but, having exposed so many

spurious effects of the earth's motion, it is rather surprising that Galileo did
not seize this opportunity to emphasize a genuine effect.

More surprising still, there is a simple variant of this experiment which
Galileo very likely could have used to demonstrate the earth's motion. In
point of fact, Tycho was right (though, of course, for the wrong reasons) in
his claim that a stone dropped from a high tower would not fall at the foot.
The stone would not be left behind, however, as Tycho thought, but would
fly on ahead (would suffer an eastward deflection), and for precisely the same
reason that the cannon ball shot southward is deflected. As shown in
Figure 5b, the radius at the top of the tower exceeds that at the foot by the
height of the tower, so that, again, the linear velocity of a point at the top is
greater than that at the foot. Neglecting the curvature of the earth, the
deflection is just the difference in the linear velocities multiplied by the time
of fall. But the linear velocity of a particle moving in a circle of radius r is $r\omega$,
where $\omega = 2\pi/T$ is the angular velocity and T is the period. Hence, the
deflection is $\omega t \Delta r$, where Δr, the difference in the radii, is the height of the
tower. Now the earth's rotational period is 23 h and 56 min, or 86 160 s,
from which I compute $\omega = 0.0000729246$. Using $s = 16t^2$, an object dropped
from a height of 400 feet will have a hang time of 5 s, and so at this height,
the deflection will be $5\omega(400') = 1.75$ in. due east of the point perpendicularly below.

Now it happens that the dome of St. Peter's in Rome is about 400 feet
high, and this dome appears to have been completed in 1628. Presumably
there were workmen on top of the dome to place the last stones. They could
have dropped a plumb line to the floor and marked the point perpendicularly
below, then used a plumb line (supplemented with naked eye sighting) to fix
a rifle barrel in a vertical position. Round bullets could then have been
dropped through the barrel. One could then have computed the mean
position at which these bullets landed, and compared the distances of this
mean to the position predicted by the two competing hypotheses, the foot of
the perpendicular and the point 1.75 in. due east of it.

One might reasonably object that this experiment lacks a control: how
could one be sure the barrel was vertically placed and that systematic biases
were absent? To surmount this difficulty, one could fire one group of bullets
at the floor and merely drop (through the same barrel held in place
throughout) a second group of bullets. Since the hang time for fired bullets
would be drastically reduced, the means should be different for the two
groups (whether the rifle is perfectly vertical or not). In fact, conservatively
estimating the muzzle velocity at $200' \, s^{-1}$, the hang time of a fired bullet is

1.75 s and the predicted deflection (or distance between the mean positions for the two groups of bullets) is 1.14 in. This would have provided quite a decisive test, provided that the gun barrel was fixed very firmly in place and the equal charges of powder measured out with some care.

One can only speculate on what effects the successful performance of this experiment would have had on the subsequent course of history. (It is doubtful that Galileo's enemies in the Church would have appreciated the subtle humor of a research grant proposal to fire bullets at the floor of St. Peter's in order to demonstrate the rotation of the earth!) Still, it is *prima facie* puzzling that Galileo never (so far as I know) proposed such an experiment. His discussion of the north—south cannon shot [*Dialogue*, pp. 176f.] leaves no room for doubting that he understood the principle of the experiment (essentially, just the distinction between angular and linear velocity). On the other hand, he was mentally geared to explode absurd consequences of the earth's motion as spurious, and even comes precariously close at times [e.g., on pp. 186—187 of the *Dialogue*] to arguing, with Copernicus, that since everything participates in the earth's motion, there can be no 'sensible' effects of that motion.

For example, in discussing the claim that cannon shots to the east will carry high [*Dialogue*, pp. 180ff.] because the target is always dropping below the tangent to the earth at the point of firing, Galileo seems genuinely uncertain whether to argue that there would be no such effect or that the effect would be 'insensible'. The former position is predicated on an erroneous principle of 'circular inertia', and, in fact, Galileo says (p. 180) "just as the eastern target is continually setting because of the motion of the earth under a motionless tangent, so also the cannon for the same reason continually declines and keeps on pointing always at the same mark so that the shots carry true". The other position, which admits a slight upwards deflection, is based on the (rectilinear) law of inertia as we know it, and it is this law which Galileo uses to compute the parabolic path of a projectile in *Two New Sciences*. Much controversy surrounds the question which form of the inertia law Galileo really endorsed.

It seems to me that he did considerable wavering between the two. Certainly, he was powerfully drawn to the hypothesis that all natural motions are uniform circular. He believed (erroneously) that, were the earth's mass entirely concentrated in its center, an object dropped from a circle rotating uniformly about that center would terminate its motion in the earth's center (rather than going into a Keplerian orbit about the center). He had no proof that the path it followed would be circular, but, in *Two New Sciences*, he gives a proof that it cannot be parabolic. For the axis of the parabola would go through the earth's center, but an object moving on a parabola grows ever more distant from its axis, and hence, could not terminate its motion in the earth's center. (That is the chief reason why he is only too willing to concede that the parabolic path – based on the simplifying assumption that the horizontal component of the motion is rectilinear – is but an approximation.)

Let us now consider the argument for the uniform circular motion of a falling stone which Galileo illustrates with Figure 5b [*Dialogue*, pp. 165–166]. First off, he assumes the 'circular inertia' principle at the outset when he describes the compound motion of the stone as a rectilinear vertical motion (uniformly accelerated) superposed on a *circular* horizontal motion. This problem is rather intractable as compared with that of superposed rectilinear motions and Galileo is reduced to *conjecturing* that the resultant path is the semicircle *CIA* of Figure 5b ending in the center of the earth.

How does he know that the stone's motion along this path is uniform? He draws equally spaced radii to the points *C,F,G,H,L,D* on the circle *CD* described by the top of the tower and then he insinuates the assumption that "the points where these [radial] lines are cut by the arc of the semicircle *CIA* are the places at which the falling stone will be found at the various times" (p. 165). Since he is able to show that the corresponding arcs on the two circles, *CD* and *CIA*, cut off by the equally spaced radii are equal, and since the top of the tower moves uniformly (with the earth) on the circle *CD*, it follows that the stone's motion along *CIA* is uniform, and, moreover, he is able to conclude that the falling stone travels "not one whit more or less than if it had continued resting on the tower" (p. 166).

Unhappily, the argument is a *petitio principii*. For how can we be sure that the stone's path crosses any of the radial lines from *A* at the same instant that the tower's path crosses that line, save by *assuming* that the stone's motion along *CIA* is uniform? To be sure, the assumption is not gratuitous. A bomb remains directly under the bombsight of a uniformly moving plane throughout its descent. But 'uniformly moving' could mean either uniform rectilinear motion (with the plane varying in elevation) or uniform circular motion (with constant elevation). The latter sense gives rise to a 'circular' form of the Superposition Principle, and that is what Galileo is (not unreasonably) assuming here. From this principle it is immediate that the stone will land perpendicularly below the point from which it was dropped; there will be no eastward deflection. That fully accounts for Galileo's failure to emphasize this very 'sensible' deflection or to propose an experiment capable of detecting it.

Historians of science have belabored Galileo for what they perceive as mere failure to overcome an existing bias in favor of circular motion. But that judgment is overly harsh (and in this, I concur with Shapere (1974), pp. 104ff.). For the uniformity of the acceleration of free fall is explained as an effect of the uniform circular motion of a free-falling projectile. And there is no gainsaying the striking simplicity of the hypothesis that all unimpeded motions, terrestrial and celestial alike, are uniform circular. That Galileo was strongly drawn to this hypothesis and was able to provide a fairly convincing plausibility 'proof' on its behalf is less a mark of his 'conservatism' than of his genius.

Like Copernicus and Oresme before him, Galileo was drawn to the relativity of circular motion and was tempted to argue that no experiment could ever determine which of two spheres (e.g., the earth and the celestial sphere) was really in uniform rotational motion and which was at rest. What made him hesitate? Quite possibly, the north—south cannon shot, a differential effect of rotatory motion that follows as much on the circular inertia law as on the rectilinear inertia law. There is no similar bar to Galilean Relativity, that the laws of physics are the same in two systems in uniform translational motion with respect to each other. That is perhaps why he wavers between the one formulation of the inertia law and the other. In addition, as Shapere observes (*op. cit.*, p. 105) there is an obvious and devastating objection to the hypothesis that a falling stone's motion is uniform circular, namely, that its momentum must be the same at every point of its path!

While the uniformity of all motion is surely a most striking simplification (if it could be maintained), we should not allow its simplicity to obscure that of the rectilinear form of Galilean physics that has come down to us, for it represents a considerable advance over impetus theory. Given the parabolic law of projectile motion (based on the rectilinear inertia law), the range and flight time of a projectile is determinable as a function of its initial velocity and angle of firing, and independently of its composition, weight, density, size or shape (if we neglect air resistance). Especially striking is the prediction that a projectile fired horizontally from an elevated plane will strike the ground at the same instant as a projectile that is simultaneously dropped from the same height. Moreover, the acceleration of free fall is no longer a local constant (different for bodies of different density) but a universal constant.

While they effectively rebut physical objections to the earth's motion, none of Galileo's arguments provides any direct reason for thinking that the earth does move (with the qualifications noted above). Nonetheless, far from militating in favor of Tycho's theory, physical considerations probably sealed its fate. Tycho's picture of the heavens made little dynamic sense, and his successor, the redoubtable Kepler, was interested chiefly in the question, '*A quo moventur planetae?*'[8] Kepler's answer was: a force emanating from the sun (his '*anima motrix*'). No dynamics is readily suggested by the Tychonic picture which posits, in effect, *two* centers of motion. The simple relation between distance and period, true without exception in the Copernican system, fails in the Tychonic. Yet, Tycho's theory was rejected, not merely because it was less symmetrical or harder to visualize kinematically,[9] but because it did not suggest an answer to what had become, with the demise of Aristotelian cosmology which Tycho's own work did so much to bring about, the central question about the planets: what *causes* their motion?

APPENDIX: THE COMPLEXITIES OF COPERNICUS'S THEORY

There is no question but Copernicus's theory was more complicated than it need have been. To see this, it will help to look first at the version (Figure 6) Kepler took as his starting point, the one, he says, Copernicus would have adopted himself had he been 'more Copernican'. In Figure 6, the planet *P* moves around the sun *S* in an eccentric orbit, the motion being uniform about an equant point *A* situated symmetrically opposite the sun with respect to the center *C* of the eccentric. Such motion is known to approximate a Kepler ellipse.[10] Ptolemy's theory is obtained from Figure 6 by fixing the earth (or the point *P* in Figure 6) and sending the sun into orbit (i.e., by simple reflection of Figure 6). The planetary theory based on Figure 6 is accurate to within first order terms in the eccentricity of the Kepler ellipse.[11] The same would be true of Ptolemy's theory but for the fact that the earth's center is made the intersection of the orbital planes, producing errors in the predicted latitudinal variations of the planets. On the other hand, the earth retains a vestigial centrality in Copernicus's version as well; it is not truly heliocentric.

I have mentioned Copernicus's distaste for equants. This led him to replace the Ptolemaic construction of Figure 6 with the equivalent epicycle construction of Figure 7. Let *e* be the eccentricity of the ellipse whose semimajor axis is the radius *a* of the eccentric circle of Figure 6. Then in Figure 6, the distance from *C* to *S* is the product *ae*, just as in the associated Kepler ellipse, and *AC* = *CS*. In the equivalent construction of Figure 7, the distance from *C* to *S* is now 3/2 of *ae*, the epicyclic center *L* turns around *C* at the same rate *P*

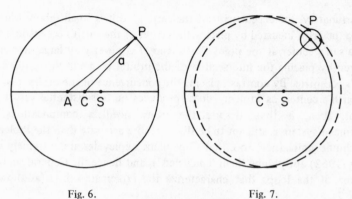

Fig. 6.   Fig. 7.

turns around A in Figure 6, and the epicycle $LP = ae/2$ turns at twice the rate of LC, the eccentric (*De Rev.*, Bk. V, ch. 4).

The equivalence of Figures 6 and 7, together with their equal numbers of fitted parameters, amply illustrates the danger of measuring simplicity by counting circles. (In both cases, the orbit of P is determined by fixing the center C, the radius a, and the eccentricity e.) While Copernicus gains by dispensing with the 'monstrous equant', he loses something by adding the extra circle of Figure 7. Nevertheless, had he applied the construction of Figure 7 uniformly to all the planets, taking the sun for the point S in each case, he would have produced a theory accurate to within first order terms in the eccentricities of the Kepler ellipses. For reasons that are not entirely clear even today, he did not do this. Instead, he made the center of the earth's orbit the center of the system, placing it at the point S in Figure 7 and making it the point in which the planes of the planetary orbits intersect. (Kepler says he did this "so as not to differ too much from Ptolemy's concept".) Moreover, in order to reconcile ancient with later Arabic and European observations, he was forced to suppose that the earth's eccentricity varies, and so he sets the earth's eccentric into a slow motion (on an epicycle carried by a sun-centered deferent). The result is that Copernicus's system is neither heliocentric nor centro-static. Having made this error, Copernicus was forced to vary the eccentricities of Venus and Mercury in a similar way and to further complicate his model of Mercury's motion.[12] Finally, the latitudinal variations of the planets can be explained heliostatically as a projection of the earth's motion, but because Copernicus has the planes of the planetary orbits intersect in the center of the earth's orbit rather than in the sun, he is led to assume a spurious variation of the orbital planes to the elliptic.

Fortunately, the eccentricity of the earth's orbit is small (about 0.0167), and so the error incurred by placing the center of the earth's eccentric instead of the sun's center at the point S of Figure 7 is never very large, nor is that incurred by placing the intersection of the orbital planes at the center of the earth's eccentric. By contrast, placing the concurrence of the orbital planes in the earth's center, as Ptolemy does, produces quite substantial errors. The upshot, then, is that, despite the many needless complications, the Copernican system is still apt to be slightly more accurate than the Ptolemaic. One should emphasize, too, that Copernicus's epicycles are extremely small (Price (1962) aptly dubs them 'epicyclets'), and hence the Copernican orbits are free of the loops that characterize the Ptolemaic orbits (as shown in Figure 7).

Gingerich (1975) compares the accuracy of the Alphosine-based ephemerides of Stoeffler and Leovitus, based on Ptolemaic theory, with that of three sixteenth century Copernican ephemerides, using modern theory to compute the actual planetary positions at the relevant times. As expected, though the differences are small overall, the Copernican ephemerides are slightly more accurate (see the tables on p. 94 of Gingerich (1975)).

A more surprising result emerges from Gingerich's study. He finds that the thirteenth century Alphonsine tables are based on a pure Ptolemaic theory with an eccentric, equant, and single epicycle per planet. Moreover, the sixteenth century ephemerides of Stoeffler agree so closely with the Alphonsine tables that Gingerich is forced to conclude that "he used the unembellished Ptolemaic system, as transmitted through the Alphonsine tables" (p. 88), and that the "complex, highly embroidered Ptolemaic system with all the added circles is a latter day myth". Commenting on Kuhn's portrayal of Copernicus's preface as "one of the classic descriptions of a crisis state", Gingerich writes (pp. 89–90):

I believe that an alternative reading is preferable. After criticizing the alternative system of homocentric spheres, and, indirectly, Ptolemy's equant, Copernicus says:

"Nor have they been able thereby to discern or deduce the principle thing – namely the design of the universe and the fixed symmetry of its parts. With them it is as though one were to gather various hands, feet, head and other members, each part excellently drawn, but not related to a single body, and since they in no way match each other, the result would be a monster rather than a man."

This 'fixed symmetry of the parts' refers to the fact that . . . the relative sizes of the planetary orbits in the Copernican system are fixed with respect to each other and can no longer be independently scaled in size. This is certainly one of the most striking unifications brought about by the Copernican system – what I would call a profound simplification. Clearly, this interlinking makes the unified man, and in contrast the individual pieces of Ptolemy's arrangement become a monster.

What has struck Copernicus is a new cosmological vision, a grand aesthetic view of the structure of the universe. If this is a response to crisis, the crisis had existed since A.D. 150. Kuhn has written that the astronomical tradition Copernicus inherited 'had finally created a monster', but the cosmological monster had been created by Ptolemy himself.

NOTES

[1] Kuhn (1975), p. 171.

[2] Dreyer (1953), p. 312. Many other conjectures have been proposed, but, one and all, they seem to me to rest on the dubious supposition that Copernicus arrived at his theory all at once and was led to it by some one consideration. I am more inclined to take Copernicus at his word (cf. the passage quoted below) and believe that ancient references to the Pythagoreans led him to experiment with the motion of the earth and that it gradually dawned on him that, not only the major irregularities of planetary motion, but the relative dimensions of the planetary orbits, were immediate consequences of this hypothesis − the 'Copernican revelation' to which the title of this chapter refers.

[3] Rheticus tells us Copernicus was led to his hypothesis by noticing the variation in the brightness of Mars (cf. the passage from Copernicus in Section 3 below). Both Copernicus and his disciple are silent about the anomalous lack of variation in the apparent size of Venus. For a thorough, if somewhat one-sided, discussion of this point and qualms about the accuracy of Galileo's telescopic observations (which removed this anomaly), cf. Feyerabend (1970), pp. 278ff.

[4] Ptolemy also assumes that the center of an eccentric orbit bisects the line joining the center of the earth to the equant point. Kepler tried to dispense with this assumption, but found that the observations demanded it (Koyré(1973), p. 186). It is a moot point, then, whether this assumption is a proper part of Ptolemaic theory or is better thought of as fitted from the observations (thereby giving the theory still another degree of freedom).

[5] See Koyré (1973), pp. 128ff. for some relevant passages from Kepler. In a similar vein, Armitage writes of Copernicus (Armitage, 1938, p. 86):

Exposed in Italy to the ideas of the Platonic revival . . . he would become familiar with a subtle and characteristically modern conception of the 'truth' of a scientific explanation. It was the ideal of that movement . . . to assign the highest degree of truth to that theory which most simply and systematically accounted for the observed facts and established the necessity of the quantitative relations observed to connect them. In the spirit of this movement, Copernicus would be disposed to embrace that theory from which followed, by mere mathematical necessity, the greatest number of celestial phenomena.

[6] This belief almost forced Copernicus to deny the physical reality of the outermost sphere of stars, for if circular motion is natural to a sphere, both the earth and the stellar sphere must have a rotation − contrary to the appearances. One also finds conceptual arguments (which resurface as the antithesis to the first antinomy in Kant's *Critique of Pure Reason*) given in *De Rev.*, Bk. I, ch. 8:

They say too that outside the heavens is no body, no space, not even void . . . . Yet surely it is strange that something can be held by nothing. Perhaps indeed it will be easier to understand this nothingness outside the heavens if we assume them to be infinite . . . .

But Copernicus then adds, in a more cautious vein:

Let us then leave to Natural Philosophers the question whether the Universe is finite or no . . . .

Although Copernicus does continue to speak of the 'sphere of fixed stars' throughout *De Rev.*, the question whether he assigned this useful fictive device any physical reality remains, to my mind, open.
[7] See Koyré (1973), p. 177 and pp. 185ff. for more on this, and Wilson (1968) for the role of Kepler's dynamical theory in his 'war on Mars'.
[8] For Kepler's dynamical speculations, cf. Koyré (1973), pp. 185–215. In my explanation of the demise of the Tychonic system, I am in essential agreement with Kuhn (1957), p. 204.
[9] Westman (1975b), esp. pp. 308–12, shows that Tycho arrived at his geoheliostatic model by inverting Copernican models, and further (p. 317) that Tycho fully appreciated the force of the Copernican arguments sketched in Sections 2–3 above, arguments which he succinctly labelled "the reason for the revival and establishment of the earth's motion".
[10] Cf. Figure 10, p. 205 of Price (1962).
[11] For the proof, see Hoyle (1962), appendix.
[12] *De Rev.*, Bk. V, ch. 25; cf. also Koyré (1973), pp. 45–47.

BIBLIOGRAPHY

Armitage, A.: 1938, *Copernicus, the Founder of Modern Astronomy*, London.
Clagett, M.: 1959, *The Science of Mechanics in the Middle Ages*, University of Wisconsin Press, Madison, Wisconsin.
Copernicus, N.: 1959, *De Revolutionibus Orbium Coelestium*, trans. by C. G. Wallis, *Great Books of the Western World*, Vol. 16, Encyclopedia Brittanica, Chicago.
Dreyer, J. L. E.: 1953, *A History of Astronomy from Thales to Kepler*, Dover, New York.
Feyerabend, P. K.: 1970, 'Problems of Empiricism, pt. II', in *The Nature and Function of Scientific Theories* (R. G. Colodny, ed.), University of Pittsburgh Press, Pittsburgh, pp. 275–353.
Galilei, G.: 1962, *Dialogue Concerning the Two Chief World Systems*, trans. by S. Drake, University of California Press, Berkeley and Los Angeles.
Galilei, G.: 1953, *Two New Sciences*, trans. by H. Crew and A. DeSalvio, Dover, New York.
Gingerich, O.: 1975, ' "Crisis" versus Aesthetic in the Copernican Revolution', in Copernicus (A. Beer and K. A. Strand, eds.), *Vistas in Astronomy* 17, 85–95.
Grant, E.: 1974, *Source Book in Medieval Science*, Harvard University Press, Cambridge, Massachusetts.
Hoyle, F.: 1962, *Astronomy*, Doubleday, Garden City, New York.
Koyré, A.: 1973, *The Astronomical Revolution*, Cornell University Press, Ithaca, N.Y.
Kuhn, T.: 1957, *The Copernican Revolution*, Harvard University Press, Cambridge, Massachusetts.
Lakatos, I. and Zahar, E.: 1967, 'Why Did Copernicus's Research Program Supersede Ptolemy's?', in Westman (1975a), pp. 354–383.
Neugebauer, O.: 1968, 'On the Planetary Theory of Copernicus', *Vistas in Astronomy* 10, 89–103.

Ptolemy: 1959, *The Almagest*, trans. by R. C. Taliaferro, *Great Books of the Western World*, Encyclopedia Brittanica, Chicago, 1959.

Price, Derek de Solla: 1962, 'Contra Copernicus', in *Critical Problems in the History of Science* (M. Clagett, ed.), Madison, Wisconsin.

Rheticus: Narratio Prima, trans. by E. Rosen, *Three Copernican Treatises*, New York.

Shapere, D.: 1974, *Galileo*, University of Chicago Press, Chicago.

Westman, R. (ed.): 1975a, *The Copernican Achievement*, University of California Press, Berkeley and Los Angeles.

Westman, R.: 1975b, 'Three Responses to the Copernican Theory: Johannes Praetorius, Tycho Brahe, and Michael Maestlin', in Westman (1975a).

Wilson, C. A.: 1968, 'Kepler's Derivation of the Elliptical Path', *Isis* 59, 5–25.

EXPLANATION

1. INTRODUCTION

Explanatory theories are more or less efficient ways of recoding data; powerful models are characterized by the dramatic reduction in the number of information units required for data storage which they effect. But, in addition, good explanatory theories are simple (Chapter 5): they impose strong constraints on the possible results of experimentation, according high probability to what was observed and low probability to what was not observed. This chapter explores the conception of explanation as coding, and also the connection between the efficiency and simplicity of a model. I begin with some remarks on the earlier literature.

2. THE COVERING LAW MODEL

The received view, the so-called 'covering law model' of Professor Hempel, equates explanation with subsumption under general laws. The presence of statistical generalizations among the covering laws serves to distinguish statistical explanations from their deductive-nomological counterparts. In his mature account of the matter[1], Hempel further distinguishes two sorts of statistical explanation, which he calls 'deductive-statistical' and 'inductive-statistical' explanation. The former comprises the subsumption of statistical regularities under broader statistical regularities, as when one derives a mean learning curve from probabilistic assumptions about the nature of the learning process. This category is also meant to incorporate the derivation of approximative laws, like the Boyle-Charles law, from probabilistic assumptions like those of the kinetic theory. We would stress the purely theoretical character of such explanations. They are really derivations of laws from laws – 'pure mathematics' – and in no sense do they account for experimental data. It is only in what Hempel calls 'inductive-statistical explanation' that particular occurrences are accounted for in terms of theory. It is in this sense of explanation that a stochastic learning model accounts for the statistical features of observed subject-item sequences, or that the Mendelian model accounts for the phenotypic frequency counts exhibited in a first generation hybrid cross.

In our view, the distinction between explanation qua deduction within a theory and the explanation of experimental data — of particular occurrences, observed frequency counts, etc. — is far more germane to the logic of explanation than is the Hempelian distinction between deductive-nomological and statistical explanation. The reason, most baldly stated, is that no experimental datum is ever a logical consequence of a model, whether that model be stochastic or deterministic. For, in practice, the free parameters of a model must be estimated before predictions can be generated.[2] In the face of discordant observations, the experimenter is always faced with the option whether to revise parameter estimates or the assumptions of the model. (Here we are ignoring the standard role of statistical analysis in determining whether departures from theoretical expectation can be ascribed to chance fluctuations or sampling errors.) To take a trivial case, a long run of heads may impugn either an estimate of the binomial parameter or the assumptions of independence and stationarity which define the binomial model. The ambivalence in question is quite ubiquitous and does not depend on the deterministic or stochastic character of the covering model. It helps to delineate a sense in which all explanation is essentially probabilistic. Insofar as a model embodies free parameters, the probabilistic character of the estimates supplied will be reflected in predictions based on the model.

The idea that experimental data can be logically consequent upon a theoretical model has played an invidious role in the philosophy of science. Accounts erected on this basis render the intellectual processes by which theories are compared and revised in the light of data wholly mysterious. From a purely deductive point of view, a discrepant observation impugns the totality of theoretical assumptions, parameter estimates, and initial conditions upon which it was predicated, without supplying the slightest hint as to which portions of the apparatus are defective, or suggesting appropriate modifications. The matter is quite otherwise when the connection between the model-cum-parameter estimates and the data is seen to be probabilistic. For then defective assumptions are recognizable as those to whose modification the probability of the obtained datum is sensitive, and the appropriate modifications, other things being equal, are those which maximize this probability.

Deductive explanation, in our sense, is prima facie non-problematic. Such qualification notwithstanding, the paramount task remains that of analyzing the modes of probabilistic argument involved in the explanation of data. The analysis we offer retains the spirit of the covering law approach while avoiding some pitfalls of Hempel's own formulation. The discussion will also throw light on the relation between explanation and confirmation.

3. THE CONTEXTUALITY OF EXPLANATION

All explanations partake in varying degrees of the core sense of 'rendering more comprehensible'. However, there are many ways, depending on the context, of rendering the obscure less obscure. One might illustrate this by listing the great diversity of partial synonyms of 'explain'. Some of these have acquired a technical sense in the literature. For example, 'explicate', in the writings of Carnap and others, has come to signify the replacement of a pre-scientific notion by an exact concept which preserves many paradigm cases of the original while parsing borderline cases so as best to serve the needs of fruitful theoretical development.

Hempel, following a host of earlier writers, has attempted similarly to isolate a technical sense of 'explanation', viz. that of subsumption under general laws. Such 'scientific explanations' constitute redescription of an event in terms of an entrenched theory. However, considerable ambiguity surrounds the latter component inasmuch as one's interest is often directed toward comparing the ability of rival theories to account for pertinent effects without regard to the evidential support they may individually enjoy. This is particularly true during the intial stages of research which are characterized by a paucity of definitive data. More often, though, one's judgments as to the cogency of rival explanations depend as well on the plausibility of the explanatory hypotheses to which appeal is made. It is this additional dimension that really distinguishes explanation from confirmation.

In what follows we shall be concerned with the logic and cogency of scientific explanations, distinguishing these from the various non-scientific modes of illumination. The latter catch-all category might incorporate such activities as elucidating an obscure line of poetry, providing directions for the operation of a digital computer, or instructing a class in subtleties of Baroque architecture. Notice that in such contexts the explanandum – if indeed it makes sense to speak here of an explanandum – is never an individual occurrence or an experimental datum. Rather, it is a meaning to be unravelled, a complex operation to be mastered, and so on. Scientific explanations, on the other hand, involve non-trivial recoding of descriptions of individual occurrences or experimental outcomes in general terms.

However, we shall not attempt to lend greater precision to this distinction, nor to the rather vague notion of recoding upon which it turns.

Obviously, not all purported explanations are recognized as such by the scientific community. The reason is that entrenched theories not only condition one's judgment as to what research problems are significant and what are the appropriate techniques for attacking them (the very design of apparatus being mediated by theory), but also yield criteria of what constitutes a genuinely scientific explanation. The impact of mechanics on methodology is well-attested, as in the following passage from Kuhn (1962), p. 102:

Before Newton was born the 'new science' of the century had at last succeeded in rejecting Aristotelian and scholastic explanations expressed in terms of the essences of material bodies. To say that a stone fell because its 'nature' drove it toward the center of the universe had been made to look a mere tautological word-play, something it had not previously been. Henceforth the entire flux of sensory appearances, including color, taste, and even weight, was to be explained in terms of the size, shape, position, and motion of the elementary corpuscles of base matter. The attribution of other qualities to the elementary atoms was a resort to the occult and therefore out of bounds for science.

Thus, the choice of terminal link in an explanatory chain will reflect theoretical commitments of the moment. Where no theoretical paradigm has emerged, concern over what constitutes genuine explanation, as well as over what stands in need of explanation, will be widespread. Such concern permeates a recent discussion of the regularities characterizing human capacities for information storage.[3]

Some writers have, I believe, taken the view that the behavioural regularities expressed in terms of the information measures themselves constitute a theory, but I am inclined to class them simply as empirical generalizations requiring theoretical analysis. It is often not easy to know when a particular relation stemming from experiments should be considered a generalization in need of explanation and when it should be introduced as an unanalyzed assumption of a theoretical system, but two features of these relations lead me to class them as generalizations. First, they are not really simple, certainly not in the sense that linearity, additivity, and independence assumptions are simple. Second, they are statements about averages – not just averages of data, which are often used as estimates of probabilities, but averages over distinct classes of responses – and so the observed regularities are bound to mask much of the possibly interesting fine detail of the behaviour.

One can only conclude that the conditions under which redescription constitutes genuine explanation – as opposed to 'mere description' – relate in part to the entrenchment and simplicity of the theories in terms of which the description is formulated. While controversy appears at times to revolve around distinct conceptions of the logic of explanation, as in disputes surrounding functional and teleological explanation, closer inspection reveals that disagreement as to what theory should guide research is more germane. Thus, we would construe functional and teleological patterns of explanation as appealing to a different species of covering law rather than as embodying a *sui generis* logic.

That explanations are typically elliptical, some explanatory premises being suppressed, marks a more superficial sense in which explanation is contextual. In some cases a theory is clearly presupposed and one is called upon to specify initial conditions. In other cases one seeks theoretical assumptions capable of mediating inferences from well-known inputs to observed outputs. This is theory construction proper. The context, theoretical and practical, determines what features of the phenomena constitute explananda.

Sometimes great care is required to specify precisely what aspects of the data call for explanation – witness the ambiguity of the question 'Why did the coin show heads on nine of the ten tosses?' Any of the following could constitute an appropriate explanans, given a suitable context: (i) a physical description of the coin and the flipping mechanism, (ii) a binomial model with biased probability of heads, (iii) a model positing after-effect. In practice, however, the question would usually be interpreted as calling for the resolution of a minor paradox, and the appropriate 'explanans' would state simply that rare events are not excluded by the laws of chance. This is an instance of a kind of explanation that is particularly pervasive and important; viz., explaining away an apparent conflict between theory and data.

4. EXPLANATION AND CONFIRMATION AS INVERSE RELATIONS

In what are ordinarily accounted scientific explanations, the explanandum is a probabilistic consequence of the explanans. (We define this notion precisely below.) Indeed, explanation is often intuitively construed as a species of hypothetical deduction: to explain is to find hypotheses from which the observed effect may be inferred with high probability. Thus to cite a familiar example, that the patient recovered from infection is explained by the fact that he was injected, i.e., his recovery was highly probable conditional on his receiving the injection. Moreover, in this example at least, it is plausible to maintain that the patient's recovery could have been predicted beforehand on the basis of the diagnosis (say bacterial infection) and the prescription (penicillin).

Hempel's formulation[4] of the argument implicit here runs as follows. Let E be the observed effect (recovery) and let H be the explanatory hypothesis (the injection). Then, according to Hempel, for the explanation to be sound E must be practically certain relative to the premisses 'H' and '$P(E/H) = 1 - \phi$', where ϕ is suitably small. In diagrammatic form:

$$\frac{\text{Patient injected} \atop P(\text{recovery/injected}) = 1 - \phi}{\text{Patient recovers}} \text{[makes practically certain]}$$

One immediate objection to this scheme is that it is insufficiently general, for in the case where the suppressed premisses are the assumptions of a model, the explanatory hypotheses will include estimates of the undetermined parameters, and will therefore assume the form '$P(H) = q$' rather than 'H' (a categorical assertion). In an informal aside, Hempel remarks that "the explanatory premisses are – however tentatively – asserted as true". But this hardly constitutes analysis of the general case where the premisses themselves are only more or less probable.

A far more serious objection is that Hempel's formulation requires a cogent explanans to render the explanandum practically certain. This underlies Hempel's support of the structural identity thesis, according to which a satisfactory explanans could have served as a basis for antecedent prediction of the explanandum. Indirectly the formulation suggests that the cogency of an explanation be measured by the magnitude of $P(E/H)$, the probability of the explanandum conditional on the explanans.

Quite apart from the vagueness of the notion of practical certainty to which it appeals, one can offer the following straightforward counterexample to the first of these claims.[5] A small number of people who are injected with penicillin, say 2%, develop an allergic skin reaction. Suppose that a patient develops such a reaction, and that it is explained by his having received the penicillin. Is the explanation unsound?|I think not! Rather, the example demonstrates that the cogency of an explanation does not depend solely on the strength of the statistical correlation in question.

The structure of the example is of an effect which has only one possible cause which nevertheless occasions the effect infrequently. Fortunately, the more searching analysis of the probabilistic argumentation involved in explanation indicated by this example is not far to seek. The example makes clear that a high probability by itself means nothing: to explain an effect E, an hypothesis H must lend E a significantly higher probability than it possesses relative to the rest of our background knowledge. E.g., smoking explains (causes) lung cancer not because smokers are practically certain to contract the disease, but because the probability that a randomly drawn smoker will contract carcinoma is appreciably higher than the probability that a randomly drawn non-smoker will.[6]

In Bayesian terms this can be expressed by saying that the *expectedness* of E is exceeded by the value which its *likelihood* assumes at the hypothesis H, in symbols:

(1) $P(E) < P(E/H)$.

By Bayes' theorem, this assertion is equivalent to

(2) $P(H) < P(H/E)$,

that is, to the positivity of the *plausibility increment* $P(H/E) - P(H)$.

It is not unnatural to identify the metric concept of degree of *confirmation* (*DC*) with the plausibility increment, writing

(3) $DC(H/E) = P(H/E) - P(H)$

for the degree to which E confirms H. I will refer to all conventional variants of the plausibility increment indiscriminately as 'the Bayesian concept of confirmation'. Positive values of (3) correspond to confirming data, negative values to disconfirming data, and zero values to neutral data. The resulting confirmation relation is both symmetric and nontransitive, so that, in particular, the so-called 'Consequence Condition' fails.[7] The Bayesian concept captures what I take to be the root sense of 'confirmation' as 'increasing the initial presumption for'.

Whenever E confirms H, we will also say that E is a *probabilistic consequence* of H. Thus, on our treatment, the dictum according to which hypotheses are confirmed by their consequences becomes definitional (for a broader sense of 'consequence'). We can now remove the vagueness inherent in Hempel's formulation by saying that, in a sound *explanation, the explanandum is a probabilistic consequence of the explanans.* (Of course a residue of vagueness still remains insofar as cogency requires that the explanatory premisses be well attested.)

The present analysis clearly excludes the counterexample to Hempel's formulation. While the antecedent injection of penicillin does not serve as a basis for predicting the rash, it does supply the bulk of the unconditional probability attaching to the latter, so that, in effect, *rival accounts have been eliminated*. Collins (1966), p. 135f., offers a similar rationale:

We think having had the penicillin explains a man's rash although penicillin seldom produces the rash. This is sensible because if it was not the penicillin it must have been something else, and the likelihood that other factors were present that might have produced the rash is not great .... Knowing that the event we want to explain *has occurred* and that this sort of event customarily only comes about in one of certain roughly known ways, we are pretty sure about an explanation as soon as we discover that conditions for one of those ways for getting *x* to be *G* obtained just before *x* got to be *G*. This is brought out in the contrast: While only 2 per cent of the penicillin injected are rash-getters, more than 99 per cent of those injected by penicillin who do get a rash get it as a result of the injection.

The theorem on total probability lends precision to the eliminative rationale provided here. Let me emphasize, however, that the contrast which Collins cites is already built into the analysis in terms of total probability. To say that other causal factors are absent is to say that $P(E/\overline{H})$ is small, is to say that almost all penicillin-injected rash-getters acquire the rash as the result of the injection (or some associated factor). It is not as though we required an additional statistical survey to determine what percentage of penicillin-injected rash-getters contract the rash via the penicillin.

An explanation is cogent when $P(E/H)P(H)$ is large relative to $P(E/\overline{H})P(\overline{H})$. In fact, the fallacy which generates the structural identity thesis is the mistaken supposition that explanatory cogency depends on $P(E/H)P(H)$ alone, rather than on its magnitude relative to that of the complementary quantity $P(E/\overline{H})P(\overline{H})$. The former measures predictability of E on the basis of H and can be small when the latter, and so also the explanatory cogency, is great. This is precisely what underlies the fact that the penicillin of Collins' example accounts for the rash without providing a basis for its prediction.

It is the rule, not the exception, that events can be explained *after the fact* on the basis of conditions and laws from which they could not have been predicted beforehand. This delineates the sense in which hindsight is better than foresight. It also suggests that the venerable, but erroneous, predesignation theory might have played a role in the formation of the structural identity thesis. According to this doctrine, the import of data for an hypothesis, or alternatively, the explanatory power of an hypothesis vis-à-vis given data, is decisively influenced by whether or not the hypothesis was formulated antecedent to the collection of the data. For Bayesians, the import of the data is entirely contained in the likelihood function of the outcome observed, and so does not depend on whether or not formulation of hypotheses preceded experimentation.

To be sure, explanatory power of an hypothesis is reflected in the ability of that hypothesis to account for additional (e.g. quantitative) features of the explanandum, or to explain analogous effects; that is, by the ability to render a more precise and detailed account than that initially called for. But this feature, too, is unrelated to whether those additional aspects of the data were predicted (or could have been predicted) beforehand or were recognized subsequently. Formulation of new models is almost universally attended by recognition of theoretically significant features of the available data which had been formerly overlooked.

A strictly deterministic conception of causality has also contributed to the conviction that explanation and prediction are symmetric. If causes are construed as sufficient conditions, so that whenever the cause is present the effect follows, and if explanation is construed as isolating the cause of an observed effect, then it is reasonable to suppose that what explains might also have predicted an effect. Unfortunately, this construal of causation has little bearing on current research in the social sciences. The obvious utility of a probabilistic conception of causality, such as that sketched in reference to smoking and lung cancer, underlines the inadequacy of overly stringent eliminative schemes of discriminatory analysis like Mill's methods, for these invariably equate causes with sufficient conditions.

One consequence of the asymmetry between explanation and prediction should be the exercise of greater caution in demanding predictive capacity of every proffered explanans: the ability to explain need not presuppose the ability to pre-ordain. And while the widespread preference for prospective over retrospective data is admittedly sound, it is grounded less in the validity of the structural identity thesis than in the systematic character of prospective samples. In other words, the preference is really for the kinds of *controlled experimentation* susceptible of statistical analysis. A well-designed experiment justifies more stringent statistical assumptions about the data generating process, and these lend obtained outcomes greater precision than they would otherwise enjoy. Retrospective studies can afford the advantages of, say, a stratified sample, if the records upon which they are based are sufficiently detailed and reliable. But this is seldom the case. Consequently, the opportunity for controlling sampling bias is normally much greater in a prospective study.

5. EXPLANATORY POWER AND EFFICIENT CODING

In an earlier article on explanation (Rosenkrantz, 1969), I proposed measuring the explanatory power of a theory (for an experiment) by the information the theory transmitted about that experiment. Transmitted information increases with the precision of a theory's predictions. And a theory is thought to provide a cogent explanation of a phenomenon or of given experimental findings when it accords high probability to what occurs and low probability to what does not occur. (A theory that fits everything explains nothing.) We saw (Chapter 7) that Kepler regarded the Copernican model as genuinely explanatory and the Ptolemaic model as not because the former fits *only* the observed planetary positions and irregularities. Explanatory power is thus closely related to simplicity or overdetermination, so much so, in fact, that I am now inclined to equate the two.

Additional insight is gained by thinking of theories as means of encoding data. Explanation, it is generally agreed, proceeds by something like metaphor: describing one thing in terms of another. The description of macro behavior or manifest effects in terms of micro behavior or latent substructure would today be considered the exemplar of sound explanatory procedure. (And this explodes the old idea that explanation goes by reduction to the familiar.) We want both explanans and explanandum to describe the same things while seeming to describe different things, and this encoding is part of what the word 'model' connotes.

To be efficient, such encoding must reduce the units of information utilized in storing or reproducing the data. To take a simple example, Bode's law

$$d_i = [3(i-1) + 4]/10 \, ,$$

where d_i is the distance of the ith planet in the sequence Mercury, Venus, ... in astronomical units, is a far more efficient encoding of the data than that provided by a mere list of the distances. In the same way, Kepler's laws represent a more efficient encoding than, say, Ptolemaic astronomy. The positions of the planets at all times can be calculated from Kepler's laws from initial positions given their periods, eccentricities, inclinations and the mean distance from the sun of a single planet.

One's intuition here is that the most *efficient encoding* of planetary positions is that which would allow a Turing machine, say, to compute the planetary positions in the shortest time (or from the shortest program[8]). Thus, efficiency, as measured by the length of the shortest computer program which gives the desired output, reflects both the amount of input required and the sheer mathematical complexity of the required calculations.

There is clearly a connection between the efficiency of its encoding and the simplicity or overdetermination of a theory (Post, 1959). If we qualify a law, we diminish its efficiency, but, equally, we complicate it in the sense of Chapter 5. A good theory should exploit all the redundancy present in the data, and this requirement relates directly to both efficiency and simplicity. Yet, the two notions are distinct, for as Post (1960) also observes, two polynomials of different degree, say $y = 2 + 3x$ and $y = 4 + 2x + 3x^2$, are equally simple in my sense but clearly not equally efficient: it requires a longer program to generate a locus or curve of higher degree. In practice, though, we are seldom interested in comparing particular polynomials (or particular specifications of a model), and where curve families or general cases of two models are compared, simplicity and efficiency comparisons will tend to agree. What seems worth stressing is that both efficiency and simplicity are reduced by adjunction of free parameters, and for this reason simpler or more efficient theories are alike preferable primarily because they tend to be better confirmed by conforming data.

We must be wary of making facile comparisons. At first blush, the Keplerian planetary orbits look more efficient (at least, mathematically simpler) than their perturbed Newtonian counterparts. As with the corresponding simplicity comparison, though, we must take the additional dynamic constraints of Newton's theory into account. Using his gravitation law, in fact, the orbit of a planet can be calculated from its initial position and velocity (Feynman et al. (1963), Section 9–7). Hence, far less input is needed for the Newtonian calculation. Moreover, as Feynman shows (op. cit.), the additional number of microseconds needed to include the perturbational effects of another planet in the calculation can be easily estimated. Newtonian theory also permits 'weighing' the planets (calculating masses). Efficiency comparisons are made much more difficult where one theory yields additional output from the given input.

NOTES

[1] Hempel (1965), Chapter 12.
[2] See the discussion of the blood group models in Chapter 5 for illustration.
[3] Luce (1963), pp. 170–171.
[4] Hempel (1965), pp. 381ff.
[5] The example is from Collins (1966); cf. also Scriven (1959).
[6] For a treatment of causality along probabilistic lines, cf. Suppes 1970).

[7] $P(H/E) - P(H)$ measures the 'overlap' of H and E and it is easy to find an F which overlaps E and H without E and H overlapping.
[8] Our notion of efficiency is closely related to Kolmogoroff complexity. For a thorough discussion of this notion, see Fine (1973), Chapter 5.

BIBLIOGRAPHY

Atneave, F.: 1954, 'Some Informational Aspects of Visual Perception', *Psych. Rev.*, **61**, 183–194.
Collins, A.: 1966, 'The Use of Statistics in Explanation', *Brit. J. Phil Sci.* **17**, No. 2.
Feynman, R. *et al.*: 1963, The Feynman Lectures on Physics, Vol. 1, Addison-Wesley, Reading, Mass.
Fine, T. L.: 1973, *Theories of Probability*, Academic Press, New York.
Hempel, C. G.: 1965, *Aspects of Scientific Explanation*, The Free Press, New York.
Kuhn, T. S.: 1962, *The Structure of Scientific Revolutions*, Chicago University Press, Chicago.
Luce, R. D.: 1963, 'Detection and Recognition', *Handbook of Mathematical Psychology*, Vol. 1 (R. D. Luce, R. R. Bush, E. Galanter, eds.), Wiley, New York, Chapter 3.
Post, H. R.: 1960, 'Simplicity in Scientific Theories', *Brit. J. Phil. Sci.* **11**, 32–41.
Rosenkrantz, R. D.: 1969, 'On Explanation', *Synthese* **20**, 335–370.
Scriven, M.: 1959, 'Explanation and Prediction in Evolutionary Theory', *Science* **130**, 477–482 (reprinted in R. Munson (ed.), *Man and Nature*, Dell, N.Y.).
Suppes, P.: 1970, *A Probabilistic Theory of Causality*, North-Holland, Amsterdam.

PART THREE

STATISTICAL DECISION

CHAPTER 9

SUPPORT

1. INTRODUCTION

Standard approaches to hypothesis testing presuppose a partition of mutually exclusive and jointly exhaustive hypotheses. It is implicit in these approaches that one can assess an hypothesis only relative to a specified set of alternatives. R. A. Fisher was perhaps the one notable exception to this rule among influential statisticians. He regarded the likelihood function as the appropriate vehicle for comparing hypotheses, but developed significance tests as the appropriate measure of evidence when no alternatives are in question.[1] Significance tests, though variously interpreted, have played a major role in research. That is unsurprising, for in the preliminary stages of an investigation – the exploratory phase – no theories of the phenomena are at hand. The researcher wishes merely to get a feel for whether a hunch or proto-theory represents a promising line of attack on a problem. Unfortunately, Fisher's conception is shot full of difficulties,[2] so much so, that many have doubted whether significance tests are more helpful than misleading.[3]

2. FISHER'S NEGATIVE WAY

Fisher believed experiments could only confute hypotheses, never confirm them. He even says (Fisher, 1935, p. 19) "every experiment may be said to exist only in order to give the facts a chance of disproving the null hypothesis". Now the null hypothesis is not ordinarily the positive hunch or proto-theory one wishes to appraise, but an opposing hypothesis of chance, denying a difference in treatments, a causal relation, or an association between traits. Yet, in practice, when a null hypothesis is rejected, that is commonly interpreted as evidence in favor of the alternative hypothesis of interest. In genetics, for example, if the data from a mating experiment serve to reject the null hypothesis of independent assortment at the pre-assigned significance level, that is taken as evidence of linkage, and one then proceeds to estimate the degree of linkage.[4] Obviously one can never rule out the possibility that the source of the observed deviation from expectations founded on the null hypothesis is something entirely unsuspected. Independent assortment may fail, for example, because the separate factors are not segregating in the expected Mendelian ratios, or for any number of other reasons. One cannot hope to exclude every such possibility or even to extricate each one for separate testing. It seems desirable, therefore, to insist that, to genuinely support the hypothesis of interest, the data actually do agree with it (and not merely disagree with the opposing null hypothesis). Before developing this idea, we must first consider the conditions under which an experimental outcome suffices to reject a null hypothesis.

3. FISHER'S PRACTICE

Fisher's precept is best gathered from his practice. He chooses an experimental random variable or test statistic by which to order the possible outcomes as to their agreement *with the hypothesis of interest*. He then finds the sampling distribution of this statistic conditional on the null hypothesis. The null hypothesis is then rejected if the observed value of the test statistic falls in one of the tails of its sampling distribution. (This description is deliberately ambiguous as between a 'one-tailed' and a 'two-tailed' test.)

The celebrated example of the tea-tasting Lady serves to illustrate (Fisher, 1935, Chapter II). A Lady claims that by tasting she can tell whether milk or tea was infused first in a cup containing a mixture of the two. Fisher's experimental design calls for the Lady to taste and classify eight cups, four having received each treatment. The test statistic is the number of correctly classified milk-first cups, call it R. (Hence the number of correctly classified cups is taken to measure her ability.) The sampling distribution of R on the null hypothesis that the Lady is merely guessing is the hypergeometric:

$$(3.1) \qquad P(R = r) = \binom{4}{r} \binom{4}{4-r} \Big/ \binom{8}{4}.$$

The null hypothesis of no ability is then 'rejected' if r is 'too large', or better, if the probability of observing r or more correct classifications is 'too small'.

How small is small enough? There is evidently no sharp cut-off and several rather different attitudes can and have been adopted. One can force the problem into a decision theoretic mold, requiring that the experiment issue in a decision to accept or a decision to reject the null hypothesis, a rule of rejection being laid down in advance. Such a rule fixes how small is small enough, and the rule is chosen on the basis of the error probabilities to which it leads (viz. the probability of mistakenly rejecting the null hypothesis when it is true in favor of some specified alternative or mistakenly accepting the null hypothesis when it is false). Neyman-Pearson theory and cognitive decision theory[5] both adduce additional criteria for selecting a rejection rule. One could adopt a very different attitude, however, and regard the probability of equalling or exceeding the observed number of correct classifications as a measure of the evidence against the null hypothesis. Fisher's own position on the matter is ambivalent to say the least.[6]

4. THE POSITIVE WAY

Let us pursue the evidential interpretation of significance tests. Fisher's conception presupposes an intuitive rank ordering of experimental outcomes as agreeing more or less well with the hypothesis of interest. The evidential interpretation implicit in Fisher's writings then comes to this: the smaller the probability (on the null hypothesis) of observing agreement with the alternative hypothesis as good as or better than that obtained, the stronger the evidence against the null hypothesis. But it seems perverse under these circumstances not to add: 'and the stronger the evidence in favor of the alternative hypothesis'. I call this interpretation 'the positive way'.

Fisher rightly insists that a suitable null hypothesis ought to be part of the analysis of every experiment. But, on the positive interpretation we are proposing, the null hypothesis no longer enters as the focus of the test, but instead, as a fixed point of comparison by which to assess the improbability of a theory's accuracy. Thus, in Fisher's example, it is not enough that the Lady score a high percentage of correct classifications; she must score an *improbably* high percentage, where improbability is measured against the null hypothesis that she is merely guessing.[7]

In general, I call the chance probability of agreement as good as that observed the *observed sample coverage* (or *OSC*) of the hypothesis or theory of interest. Here 'chance probability' means 'probability conditional on a suitable null hypothesis of chance'. The latter might typically assert that the universe is maximally disorderly (subject, perhaps, to given constraints supplied by the background knowledge). In many applications a 'maximally disorderly universe' can be taken to be one in which all experimental outcomes are equiprobable. In that event, OSC is just the proportion of possible experimental outcomes whose agreement with the theory of interest is no worse than that of the actually observed outcome, where agreement is measured by the chosen test statistical or intuitive rank ordering of outcomes.

5. ILLUSTRATIONS

The examples to follow point to two conclusions: the ready interpretability of non-parametric statistics in terms of the OSC concept, and the ease of approximating the OSC where rank order or related test statistics are chosen. Consequently, much of current statistical practice can be made sense of in terms of OSC and the 'positive way' in a manner that avoids standard objections to the 'negative way'. The reader who is uninterested in the details can pass on to Section 6.

EXAMPLE 1 (*the tea-tasting Lady*). The OSC is given by adding the hypergeometric probabilities (3.1) for those r which exceed the observed number of correct classifications. Given that this probability falls below the pre-assigned significance level, Fisher concludes that "either a rare chance has occurred or the null hypothesis is false". We conclude (much less decide) nothing, but interpret the improbability of that high a score as an index of the evidence favoring ability. Despite the difference of interpretation, comparisons of the sort Fisher cites can still be made. Thus, we can compare the evidence from different experiments (e.g., determine if one mistake with 4 cups is more indicative of ability than 3 mistakes with 16 cups), and compare the efficiency of different experimental designs (say, whether it is more efficient to assign unequal numbers of cups to the two treatments).

EXAMPLE 2 (*randomness*). Often we wish to test an assumption of randomness or independence against a trend alternative. In the furor some years back over the U.S. draft lottery, for example, it was claimed that the lottery was not truly random, but that dates later in the year had a tendency to be drawn earlier. Since capsules containing these dates were inserted last, one would expect just this tendency if shuffling were inadequate. (The capsules inserted first would tend to sink to the bottom and be drawn last.) To assess a trend alternative, we need an appropriate test statistic. Order or rank statistics are obvious candidates, being scale invariant. For each month, starting with January, we sum the trial numbers on which the dates of that month were drawn, and then rank the sums from highest to lowest, writing x_i for the rank of the ith month, $i = 1, 2, \ldots, 12$. Several related functions of the ranks provide a useful test statistic, among them Kendall's tau and Spearman's rho.[8] Both have approximately normal sampling distributions, and consequently, an approximate OSC can be computed using either one. We illustrate this for Kendall's tau, which is defined by

$$(5.1) \qquad \tau = 2S/n(n-1),$$

where $S = P - Q, P$ and Q being calculated as follows: for each rank X_i, we count the number of larger (resp. smaller) ranks to its right, summing these counts over $i = 1, \ldots, n$; the first sum is P and the second is Q. Hence, $P + Q = n(n-1)/2$, the number of ways of choosing 2 of the n items for comparison. Hence, Q measures the number of items which are ranked differently in the two orderings $1, 2, \ldots, n$ and $X_1, X_2, \ldots, X_n$, or equivalently, Q is the minimum number of transpositions of adjacent items needed to bring the latter into coincidence with the former. When the two rankings coincide, $S = P$, $Q = 0$, and so $S = n(n-1)/2$, its maximum, and $\tau = 1$. At the other extreme, when one ranking is the inversion of the other, $S = Q$ and $\tau = -1$. Hence tau is a normalization of S, and ranges between -1 and $+1$. The sampling distributions of S can be calculated from the obvious recursive relation $u(n, s) = u(n, s - n) + u(n, s - n + 2) + \cdots + u(n, 0) + u(n, 2) + \cdots + u(n, s + n)$, where $u(n, s)$ is the number of rankings of n items for which $S = s$. The distribution is symmetric about 0, with mean and variance given by:

$$(5.2) \qquad E(S) = 0, \text{var}(S) = n(n-1)(2n+5)/18 \,.$$

It is tabulated for $n = 4, 5, \ldots, 10$ in Kendall (1970), Appendix 1, and for larger n, the normal approximation may be used. Suppose that for $n = 12$, the observed ranks (starting with January, from highest to lowest) were:

$$1, 3, 2, 6, 5, 4, 8, 7, 9, 11, 12, 10.$$

Then $P = 59$, $Q = 7$, $S = P - Q = 52$, and $\tau = 52/66 = 0.7879$. By the normal approximation, $P(S \geqslant 52) = 1 - \Phi((52 - 1)/14.583) = 0.000233$. (Since the possible values of S increase by 2, we subtract 1 from 52 in the continuity correction, rather than 0.5.) The chance probability of agreement this good with the suspected trend is thus less than 3 in 10 000. The actual ranks in the U.S. draft lottery were:

$$1, 2, 3, 4, 5, 6, 7, 8, 9, 10, 11 \,,$$

giving $S = 64$, $\tau = 64/66 = 0.9697$, and an approximate OSC of 0.000009, or less than 1 in 100 000. I.e., with a truly random lottery, agreement with the trend alternative this good or better would be expected in fewer than 1 in every 100 000 repetitions of the experiment!

EXAMPLE 3 (*comparative experiments*). Testing the equality of two means, as when we wish to compare two treatments, or a treatment with a control, is surely a basic problem of statistics. We assume only the continuity of the underlying distributions F and G, and wish to test the hypothesis that one of the populations is stochastically smaller than the other; i.e., $F(x) \leqslant G(x)$ for all x. As in Example 2, this is certainly a very heterogeneous family of hypotheses, and there would be little point in attempting to compute a complete likelihood function or power function unless we had definite information about the form of the underlying distributions. Lacking such information, the OSC is a useful surrogate. To obtain a scale invariant test statistic, rank the s controls and t treated subjects according to their scores on the quantitative variable of study. Then both the ranks themselves and the rank sums W_C (for controls) and W_T (for treated) are scale invariant. Clearly, $W_C + W_T = n(n + 1)/2$, where $n = s + t$ is the sample size, and so it doesn't matter which of W_C or W_T is chosen, since each determines the other. The means and variances are:

(5.3) $E(W_C) = s(n + 1)/2, \qquad E(W_T) = t(n + 1)/2$,

(5.4) $\text{var}(W_C) = \text{var}(W_T) = st(n + 1)/12$.

For large n, the standardized variates are approximately unit normal. The approximation is good for $n \geqslant 12$, and for smaller values, the exact sampling distribution of the related Mann-Whitney statistics, $U_C = W_C - s(s + 1)/2$, $U_T = W_T - t(t + 1)/2$, are tabulated in Hodges and Lehmann (1970), Table H. In an example where large values of W_C are indicative of the treatment's efficacy, suppose the scores are 8.3, 7.6, 8.1, 9.0, 6.2, 7.8, 8.2 for $t = 7$ treated, and 5.4, 7.1, 6.8, 8.4, 6.1 for $s = 5$ controls, giving $W_C = 42$. By consulting the table in Hodges and Lehmann, the exact OSC is $P(W_C \geqslant 42) = 59/792 = 0.074$. On the other hand, $E(W_C) = 32.5$, $\text{var}(W_C) = 37.9167$, and the normal approximation gives 0.072.

EXAMPLE 4 (*association*). There are many measures of the association of traits applicable to contingency tables. The normal approximation is again available for computing the OSC, but I must forego what would be a lengthy discussion of the details and refer the interested reader to the masterly presentation of the subject in the three-part series of articles by Goodman and Kruskal cited in the references.

The OSC allows one to attest hypotheses without reference to alternatives, or rather with regard to a fixed chance alternative which asserts the universe is, in the relevant respect, maximally disorderly. We are then in a position to compare non-exclusive hypotheses, even hypotheses drawn from different areas of science, by comparing the improbability of their accuracy. Especially where the respective experiments have the same probability distribution, say multinomial, such comparisons can be quite meaningful.

Finally, we can amalgamate data from several experiments to obtain an overall appraisal of support. To do this we must overcome the apparent difficulty of comparing a case where one experiment of a pair fits the theory very well and the other very poorly with a case where both experiments of a pair fit the theory moderately well. We propose the following. First, for the case of independent experiments, the OSC of the composite experiment with these experiments as factors, is defined as the product of their individual OSC's. Secondly, for dependent experiments, we again compute the product of the OSC's at the observed outcomes of the component experiments and count a possible composite outcome as fitting the theory as well when the product of the OSC's of the outcomes of its components does not exceed the product for the given observed composite outcome (just as in the case of independent experiments). Only now, the probability of a composite outcome is computed on the assumption of dependence, and hence, it can never be smaller than the like probability for independent experiments. In this way our proposal accounts for the higher support which a theory receives from independent experiments, other things being equal.

6. INTERPRETATION

It is time to look more closely at the interpretation of observed sample coverage, qua index of support. Clearly OSC decreases (and support increases) with the accuracy of the theory, the size or precision of the experiment, and the sample coverage of the theory. By the *sample coverage* of a theory with respect to a given experiment, I mean the chance probability that the outcome of the experiment will 'fit' the theory, where a criterion of fit is specified. In Chapter 5, you recall, I proposed measuring the *simplicity* of a theory (relative to an experiment) by its sample coverage. The smaller the a priori chance that the theory will fit the outcome of the contemplated experiment, the simpler the theory is in my sense.

Support tends to increase, and OSC to decrease, with the simplicity of an hypothesis. It is natural to conjecture that support and OSC are inversely related, and, more precisely, that in a comparison, the hypothesis with smaller OSC will have higher (average) likelihood. Thus we obtain a Bayesian interpretation of 'positive' significance tests *by regarding the ratio of the OSC's of two exclusive hypotheses as an approximate Bayes factor*. If this thesis (I call it the *sample coverage rule*) is sound, at least as a rule-of-thumb, then it justifies us in viewing the OSC as a natural extension of, or approximation to, Bayesian methods where the likelihood function cannot be computed (whether for theoretical or practical reasons).

The sample coverage rule, like the simplicity thesis, is not without exceptions[9], but again, these appear to be limited almost exclusively to equivocal outcomes. As an example, suppose that we wish to compare the uniform trinomial with all three category probabilities equal to 1/3 with the one-parameter trinomial with category probabilities $p/6$, $5p/6$, $1-p$, $0 \leqslant p \leqslant 1$. Consider the possible sample points (n_1, n_2, n_3) with $n_1 + n_2 = 20$, and $n = 36$ the sample size. Then $\hat{p} = 20/36$, or 4/9, is the maximum likelihood estimate of p, and hence the best-fitting case of the one-parameter model has category probabilities 2/27, 10/27, 15/27, and the expected category counts on this hypothesis are 2.67, 13.33, 20, while they are 12, 12, 12 on the uniform trinomial. As we vary $(n_1, n_2) = (n_1, 20 - n_1)$ from $(0, 20)$ to $(20, 0)$, the outcomes at first strongly favor the one-parameter model and then favor the uniform trinomial. The remarkable thing is that the transition occurs at the same cut-off point, viz., in passing from $(5, 15)$ to $(6, 14)$, whether we assess support in terms of OSC or average likelihood.

To clarify the example, it is necessary to say something about the calculation of OSC for composite hypotheses or models with free parameters (cf. Section of Chapter 5). Since chi square criteria of fit are rather suspect from a Bayesian point of view, we prefer to assess fit in terms of *direct confidence regions* (or *DCR's*), viz., regions which contain the experimental outcome with specified probability conditional on the given hypothesis. For multinomial hypotheses, using the multinormal approximation, the DCR's are $k - 1$ dimensional ellipsoids, k the number of categories, and the boundaries of these ellipsoids are contours of constant probability. Given an hypothesis with free parameters, its DCR is defined as the union of the DCR's of all its special cases, hence, in the multinomial case, as the envelope of a family of ellipsoids. (In our trinomial example, with k = 3, the DCR's are ellipses.)

7. CONCLUSION

I want briefly to indicate how a number of puzzles that bedevil Fisherian significance tests[10] are handled by the 'positive way'. The first concerns the difference between substantive and statistical significance, and the related point that, paradoxically, significance with a small sample size constitutes stronger evidence against a null hypothesis than significance with a large sample size.

In practice, we want to know whether an observed deviation from null expectation is large enough to suggest a true deviation that is of material importance or scientific interest. Hodges and Lehmann (1954) propose tackling this problem by testing an approximate null hypothesis, viz., the hypothesis that the parameter falls within a specified neighborhood of the null value, values outside this neighborhood being those which are deemed substantively significant.[11] This proposal runs up against the difficulty that theoretical considerations do sometimes single out a particular value of a parameter. (Mendelian genetics abounds in such cases.)

Consider next the puzzle about large samples. Suppose we are testing the sharp null hypothesis $p = 0.5$ against $p \neq 0.5$. As the sample size is increased, an orthodox significance test becomes ever more sensitive to ever smaller departures from the sharp null hypothesis. In the limit, as the sample size becomes infinite, the probability that a sharp null hypothesis will be rejected, even when it is true or as close to true as makes no difference, approaches practical certainty. In the face of this fact, Kendall and Stuart (1967)[12] advise one to reduce the significance level as the sample size is enlarged, so as to maintain a proper balance between sensitivity to true alternatives and security against erroneous rejections of the null hypothesis. But lacking any theoretically grounded rate-of-exchange between the two desiderata, security and sensitivity, their proposal strikes one as ad hoc. It is most interesting to note in this context that the average likelihood test (p. 111) does precisely what Kendall and Stuart recommend, but does it in a mathematically determinate way. One can easily check, for example, that its significance level is near 0.10 at n = 10, but drops down to 0.05 by the time n is 100.

It should be clear, finally, that increasing the sample size always has positive value if our interest is in the support *for* an hypothesis. Berkson (1942) considers a test of a physician's ability to determine the sex of a fetus in utero. In case 1, he is right in 6 of 10 trials; in case 2, in 505 of 1000 trials. The probability that he will be right at least that often by chance is the same in both cases, viz., about 0.38. "But", Berkson writes, in contrast to case 1, "the experience 2, being based on large numbers is convincing positive evidence of the truth of the null hypothesis within practical limits". Berkson is only able to account for this on vague intuitive grounds. But if we are interested in the evidence *for* the null hypothesis that the physician cannot discriminate, we see the proportion of outcomes that fit it as well is 3/11 in case 1 and 11/1001 in case 2, so that the evidence is indeed much stronger in the second case.

NOTES

[1] Cf. Anscombe (1963).
[2] Cf. Rosenkrantz (1973) and Spielman (1974).
[3] For a collective critique of significance tests from a user's point of view, cf. Morrison and Henkel (1970).
[4] A treatment of linkage from a Fisherian standpoint is contained in Mather (1951).
[5] Unbiasedness and invariance are two additional criteria to which orthodox statisticians appeal when uniformly most powerful tests do not exist; cf. Lehmann (1959), Chapters 4–6. Levi (1973) appeals instead to the content of an hypothesis, measured relative to a specified set of alternatives.
[6] As in Fisher (1956), Chapter 3.
[7] Fisher (1935), p. 18 remarks in this connection that "the rare case of 3 right and 1 wrong could not be judged significantly merely because it was rare seeing that a higher degree of success would frequently have been scored by chance".
[8] For a very thorough treatment of rank correlation statistics, cf. Kendall (1962).
[9] The exceptions to the simplicity thesis noted in Chapter 5 also serve as exceptions to the sample coverage rule. E.g., in comparing $0.49 \leqslant p \leqslant 0.51$ and $p < 0.49$ or $p > 0.51$, the OSC of the former is clearly smaller when n of $2n$ trials issue in 'success', even though, as we have seen, the average likelihood of the latter (more complicated) hypothesis is higher at modest sample sizes. Even where the sample size is large enough to discriminate the hypotheses, exceptions might still occur when the outcome is 'equivocal' in the sense of fitting neither hypothesis of the comparison at all well; cf. Section 6 of Rosenkrantz (1976).
[10] Morrison and Henkel, *op. cit.*
[11] If both kinds of error are considered equally serious, then the Neyman-Pearson theorist is hard put to say what is a best test of $p = 0.5$ vs $p \neq 0.5$. But if he tests instead the approximate null hypothesis $p_1 \leqslant p \leqslant p_2$ against $p < p_1$ or $p > p_2$, then he is able to say the optimal test is that which equates the significance level with the power of the test against p_1 and p_2 (neither error being considered more serious than the other). That, from an orthodox standpoint, is another virtue of testing an approximate null hypothesis. I owe this observation to R. N. Giere; his considered view appears to be that, in practice, we never have a truly sharp null hypothesis anyway.

[12] They write (p. 183): "The hypothesis tested will only be rejected with probability near 1 if we keep α fixed as n increases. There is no reason to do this: we can determine α any way we please, and it is rational . . . to apply the gain in sensitivity arising from increased sample size to the reduction of α as well as of β".

BIBLIOGRAPHY

Anscombe, F. J.: 1963, 'Tests of Goodness of Fit', *J. Roy. Stat. Soc., B.* **25**, 81–94.

Berkson, J.: 1942, 'Tests of Significance Considered as Evidence', *J. Amer. Stat. Assoc.* **37**, 325–335.

Birnbaum, Allan: 1969, 'Concepts of Statistical Evidence', in *Philosophy, Science and Method* (ed. by S. Morgenbesser, P. Suppes, and M. White), St. Martin's Press, New York.

Fisher, R. A.: 1935, *The Design of Experiments*, Oliver and Boyd, Edinburgh.

Fisher, R. A.: 1956, *Statistical Methods and Scientific Inference*, Oliver and Boyd, Edinburgh.

Goodman, L. and Kruskal, W. H.: 1954, 1959, 1963, 'Measures of Association for Cross-Classification', Parts I, II, and III, *J. Amer. Stat. Assoc.* **49**, 732; **54**, 123; **58**, 310.

Kendall, M. G.: 1962, *Rank Correlation Methods*, Griffin, London.

Kendall, M. G. and Stuart, A.: 1967, *The Advanced Theory of Statistics*, Griffin, London.

Laddaga, R.: 1973, *A Defense of the Likelihood Principle*, Master's Thesis, University of South Carolina.

Lehmann, E.: 1959, *Testing Statistical Hypotheses*, Wiley, New York.

Mather, K.: 1951, *The Measurement of Linkage in Heredity*, Methuen, London.

Morrison, D. F. and Henkel, R. E.: 1970, *The Significance Test Controversy*, Aldine, Chicago.

Rosenkrantz, R. D.: 1973, 'The Significance Test Controversy', *Synthese* **26** 304–321.

Rosenkrantz, R. D.: 1976, 'Simplicity', *Foundations of Probability and Statistics and Statistical Theories of Science* (ed. by W. Harper and C. Hooker), Vol. 1, D. Reidel, Dordrecht, Holland, 167–203.

Spielman, S.: 1974, 'The Logic of Tests of Significance', *Phil. Sci.* **41**, 211–226.

CHAPTER 10

TESTING

1. INTRODUCTION

Interpreted as indices of the support for an hypothesis, Fisherian significance tests are assimilable into the corpus of Bayesian methods. The observed sample coverage is an approximation to the average likelihood, and one that is often more convenient to use (or which can be used as a surrogate when the likelihood function cannot be computed). Moreover, the OSC has a clear and definite meaning as a measure of the improbability of a theory's accuracy. Any two theories or hypotheses can be compared in this respect, whether they are exclusive or not, and whether they are drawn from the same or disparate fields of science. In what follows, I will refer to this interpretation of significance tests as 'the Bayesian evidential interpretation'.

2. INDUCTIVE BEHAVIOR

I have alluded to another non-evidential or behavioral interpretation of significance tests, one which assimilates the testing of hypotheses to the making of decisions. That is, a *test* is expressly conceived as a *rule of decision* (or *decision function*), viz. a function which associates with each outcome of a predesignated experiment a decision to 'accept' or to 'reject' the tested hypothesis. (The tested hypothesis and the alternatives must also be specified in advance.)

This approach to testing, developed largely by Jerzy Neyman, E. S. Pearson, and Abraham Wald, eschews all reference to 'inductive inference'. The question, what is the best inference from an observed sample to an incompletely known population, is regarded as lacking any clear and clearly defensible answer (see below). Neyman (1957), p. 15, writes:

... the common element of all writings on the inductive reasoning approach appears to indicate the conviction of the authors that it is possible to devise a formula of universal validity which can serve as a normal regulator of our beliefs. Furthermore, the absence or, at least, the rarity of discussions of consequences of this or that choice of action suggests another conviction, namely that human actions are motivated predominantly, if not exclusively, by beliefs.

Of the competing formulae, Neyman mentions that of Bayes, Fisher's use of likelihood (without priors), of significance levels (evidentially interpreted) and fiducial inference. But at least much of the diversity strikes me as apparent. No one, to my knowledge, has seriously questioned the appropriateness of Bayes' rule for revising probabilities. And we have seen how to assimilate Fisherian significance tests into this framework.

The difficulty lies, if anywhere, in the specification of the priors – the probability inputs needed to generate probability outputs from Bayes' rule. Objectivist Bayesians, like E. T. Jaynes and Sir Harold Jeffreys, have proposed normative rules for selecting a prior, most notably, the maximum entropy rule discussed in Chapter 3, using invariance requirements as prior constraints where appropriate. These rules are proposed, of course, in a tentative spirit, and the criticisms that have been lodged against them will have to be weighed very carefully (cf. Chapter 3). And yet, Neyman's remark (*op. cit.*, p. 16) that "the beliefs of particular scientists are a very personal matter and it is useless to attempt to norm them by any dogmatic formulae" seems to argue past the objectivist Bayesian position, which, to quote Jaynes (1968) again, "requires that a statistical analysis should make use, not of anybody's personal opinions, but rather the specific factual data on which those opinions are based".

Above all, objectivist Bayesians reject the imputation that prior distributions are subjective or judgemental in a way that data distributions are not. As discussed in Chapter 3, the same reasoning processes used in reaching a prior are used in reaching a data distribution or a probability model of a natural phenomenon. The assumptions which underlie a prior distribution are every bit as corrigible as those which underlie a data distribution or probability model of a natural phenomenon. And, by the same token, they are as empirically confirmable in the one case as in the other. There are simply no epistemological distinctions to be drawn here.

Apart from the allegation that approaches based on 'inductive inference' force ('private' or 'personal') beliefs into a procrustean bed, the passage I quoted from Neyman contains the curious suggestion that this approach exaggerates the importance of belief as a determinant of action. On occasion, as Neyman rightly observes, we act *against* our beliefs, as when we take out flight insurance on an airplane trip.

Bayesians certainly have no inclination to downgrade the importance of 'the consequences' for practical decision making. Their insistence is rather that the inferential part of a decision problem (whose output is a probability distribution over states) be separated from its utilitarian component, embodied in a utility function over act-state pairs or consequences. There is no implication that the Bayes act will be best under the most probable state, and this suffices to show there is no necessary connection between belief and action of the kind Neyman finds objectionable. But let us delve more deeply into the relation between belief and action as it is conceived by Neyman.

His official doctrine is that a problem of testing hypotheses is *literally* a decision problem. "The content of the concept of inductive behaviour", he writes (*op. cit.*, p. 18), "is the recognition that the purpose of every piece of serious research is to provide grounds for the selection of one of several contemplated courses of action". Imagine the parameter space or set of states partitioned in such a way that each set of the partition is associated with a certain action (the 'best' action under any state in that set). The set of states is called an 'hypothesis'. To accept an hypothesis is then to choose the action associated with that hypothesis.[1] Now, under the expected utility rule, it is easy to produce multi-action decision problems (i.e., problems with three or more acts) where the Bayes act is not best under any state. That points to a limitation of the formulation just outlined in terms of 'hypotheses'. Neyman-Pearson theorists, however, have never lavished a great deal of attention on multi-action problems, and for two-action problems, the formulation in terms of hypotheses seems adequate.

Typical dichotomous comparisons that arise in science include: whether two genes are linked or not, whether the $A-B-O$ blood types are determined by a single tri-allelic gene or by two independent by-allelic genes, whether the universe is expanding or not, whether paired associate learning is incremental or all-or-nothing, and so forth. What specific practical decisions are associated with these fairly representative inferential problems? No doubt one could imagine the utility of some decision turning on the answer to any of these queries, but there is no *particular* decision problem tied to any of them.

If one wishes to maintain a thesis at any cost, one can, with Neyman (*op. cit.*, p. 10), include under the rubric of 'action' or 'decision' the "assumption of a particular attitude towards the various sets of hypotheses mentioned". Making that move, however, hardly leaves Neyman in a position to accuse Bayesians of fudging the distinction between belief and action.[2] The fundamental question, though, is whether there is anything to be gained by reporting the output of a data analysis as a decision to accept or reject an hypothesis when there is no specific action taken or contemplated. It is only by equivocating on 'consequence' that Neyman is able to disguise the otiose role of the decision theoretic formulation in treating inferential problems. It is axiomatic that rational decision be based on a consideration of the consequences of the contemplated alternatives. But where no specific actions are contemplated, there can be no 'consequences', properly so-called, and no rule for decision theory. Neyman slips this hold by invoking a novel sense of 'consequence', as in the following passage (Neyman, 1957, p. 16):

While rejecting the inductive reasoning approach because of its dogmatism, lack of clarity, and because of the absence of considerations of consequences of the various actions contemplated, I am appreciative of the several sections of the literature that this approach generated. For example, I am highly appreciative of the literature concerned with the properties of maximum likelihood estimates. However, it will be noticed that the point of view behind this literature is that of inductive behavior. In fact, the question studied is "what will the consequences of using maximum likelihood estimates be?"

That Neyman can use the word 'consequence' in two entirely unrelated senses in one paragraph goes no little way towards explaining the otherwise mysterious gap that separates orthodox precept from orthodox practice. Even as he pays lip-service to the conception of statistics as 'decision making under uncertainty', Neyman's research papers on cloud-seeding experiments, the distribution of galaxies[3], and so forth, make no mention whatever of "the consequences of the various actions contemplated" (or, indeed, of any specific actions contemplated), nor do they mention losses or discuss the relative seriousness of type I and type II errors. So far as their use of significance tests goes, they are indistinguishable from the research papers of R. A. Fisher.

At times, Neyman seems to allege only that *cognitive* decisions are taken (Neyman (1957), p. 14):

In our last publication on the subject, we more or less abandoned (an act or will!) our original model involving the hypotheses of (a) stationary universe, (b) simple clustering, and (c) a particular mechanism of observational errors, *because the agreement between the consequences of this model and the results of Shane's observations does not seem satisfactory to us.* Thus, we had to look for new avenues; after weighing the consequences of this or that possible modification of the original model, we decided (an act of will!) to give it another chance, with a modification of the postulated mechanism of observational errors. Regretfully, in order to effect this modification in definite terms, a new observational program became necessary. (My italics.)

But what consequences did Neyman and his associates weigh in 'deciding' to 'give the original model another chance' with a modified mechanism of error? The passage, and especially the portion I underscored, suggests rather that the bearing of the evidence on the original model, and more especially on the question which of its assumptions was responsible for the poor agreement with Shane's observations, was what the investigators weighed. Be that as it may, the 'decision' to modify rather than abandon the original model was taken retrospectively, it appears, and was not based on any 'rule of inductive behavior' consciously adopted in advance, with a view to such subsequent 'cognitive decisions'. That is why Neyman ends his remarks on this example by stressing the necessity of a 'new observational program' to test the modified set of assumptions.

The sense of 'consequences' now in question is not even predicable of actions, but only of decision rules. Under or over-estimating a parameter may have 'consequences' in the ordinary sense in which mis-diagnosing a patient may be said to have consequences, but that is not at all what Neyman has in mind when he refers to the consequences of using maximum likelihood estimates. He has in mind the properties of ML estimat*ors*, qua rules of estimation, above all, their asymptotic behavior. Similarly, in testing hypotheses, the 'consequences' are not consequences of particular actions based on a statistical test, but the 'performance characteristics' (e.g., error probabilities) associated with that test, qua decision rule. It is only in a Pickwickian sense that we can equate the consequences of an action with the error probabilities associated with a rule that led to that action. The consequences of convicting an innocent man or releasing an infected patient are one thing; the long-run frequencies with which a given procedure or rule commits these errors is quite another thing.

Once the two senses of 'consequence' are separated, we see that Neyman's notion of 'inductive behavior' has really little connection with decision theory *per se*. Interest is focused instead on the performance characteristics of rules of rejection for statistical hypotheses.

Divorced from action, the notions of acceptance and rejection lose all clarity. We are thrown back on the vague and fluctuating epistemological uses of 'accept *H*' which range all the way from 'being tentatively disposed to premiss *H* in theoretical deviations and practical decision', at one extreme, to 'holding *H* immune from revision', at the other extreme. In much current usage, 'acceptance' merely signals the conclusion that the 'accepted' hypothesis is in satisfactory agreement with the data, witness the following passage from one of the more authoritative expositions of orthodox theory (Kendall and Stuart, 1967, p. 163):

It is necessary to make it clear at the outset that the rather peremptory terms 'reject' and 'accept', used of a hypothesis under test ... are now conventional usage, to which we shall adhere, and are not intended to imply that any hypothesis is ever finally accepted or rejected in science. If the reader cannot overcome his philosophical dislike of these admittedly inapposite expressions, he will perhaps agree to regard them as code words, 'reject' standing for 'decide that the observations are unfavorable to' and 'accept' for the opposite. We are concerned to investigate procedures which make such decisions with calculable probabilities of error, in a sense to be explained.

Not even this passage says what its authors really intend; the 'error probabilities' mentioned in the last sentence cannot appertain to the 'decisions' cited in the penultimate sentence. An hypothesis may turn out mistaken even though the 'decision' to regard it as a good fit to the data was perfectly in order. Rather, we must equate the 'decision' whose error probability we seek with the singling out of one of several possible answers to a question, as in identifying the degree of a polynomial regression, the largest of k means, or stating whether or not two traits are associated. A *test* then associates with each possible outcome of the relevant experiment a purportedly correct answer to the question posed, where a set of mutually exclusive and jointly exhaustive alternatives is specified in advance. A test's *score* is the probability that it identifies the correct answer. An *optimal test* for the question posed maximizes the score. And, more generally, if a test has 'good' error characteristics, the answer it identifies as correct is held to be supported, or in good agreement with the data, while rejected answers are held not to be in satisfactory agreement with the data or ill-supported.

We have arrived, then, at still another interpretation of statistical tests, one that seems to inform the bulk of current orthodox practice. As in the Fisherian approach, tests are regarded as conventionally chosen criteria of satisfactory agreement between the data and tested hypothesis. But, where Fisher based the choice of a test or criterion solely on its probability of mistakenly rejecting the null hypothesis (or so it is held), the new interpretation also takes cognizance of alternative answers and the probability of accepting the null hypothesis when an alternative hypothesis actually obtains (type II error probability). A new terminology goes with the new interpretation. The probability of (correctly) rejecting the null hypothesis when some specified alternative is true is called the *power* of the test against that alternative, and the function which associates with each such alternative or parameter value the power of the test against that alternative is called the *power function* of the test. Thus, the power function measures the sensitivity of the test to various alternatives. The significance level, or type I error probability, is now called the *size* of the test. The size and power function are the *error characteristics*, and these are among the performance characteristics.

A test which, among all those of its size, has maximal power against *every* alternative to the null hypothesis, is said to be *uniformly most powerful* (or

UMP) of that size. UMP tests are clearly preferable, from this point of view, but they do not always exist.[4] Where they do not exist, even when attention is confined to tests with additional 'nice properties', one falls back on a test with 'satisfactory' performance characteristics for the problem at hand (one 'satisfices' instead of optimizing). The interpretation of a statistically significant outcome then comes to this: the outcome is in disagreement with the null hypothesis by the lights of a criterion which would have satisfactory error characteristics if applied iteratively to the problem of identifying which of the considered answers to the question posed is correct. This interpretation is also broadly inferential or evidential, though it differs from the Bayesian evidential interpretation canvassed in Chapter 9. I will refer to it as the *frequency-based evidential interpretation*, or, with greater brevity, as the *eclectic interpretation*. The following passage from Cox (1958), although it refers to confidence intervals, seems to me a representative expression of this point of view:

If we consider that the object of interval estimation is to give a rule for making on the basis of each set of data, a statement about the unknown parameter, a certain preassigned proportion of the statements to be correct in the long run, consideration of the confidence distribution may seem unnecessary and possibly invalid. The attitude taken here is that the object is to attach, on the basis of data S, a measure of uncertainty to different possible values of θ, showing what can be inferred about θ from the data. The frequency interpretation of the confidence intervals is the way by which the measure of uncertainty is given a concrete interpretation, rather than the direct object of the inference. From this point of view, it is difficult to see an objection to the consideration of many confidence statements simultaneously.

Here there is no attempt to reduce inference to decision, or to equate the two. Indeed, in Cox's view, "a statistical inference is regarded . . . as having an explicitly measured uncertainty, and this is to be thought of as an essential distinction between statistical decisions and statistical inferences" (Cox, 1958, p. 358). And to the argument that in making an inference, we are really 'deciding' to make a statement of a certain type, Cox replies (p. 359) that "one of the main problems of statistical inference consists in deciding what types of statement can usefully be made and exactly what they mean", whereas "in statistical decision theory, on the other hand, the possible decisions are considered as already specified".

The eclectic interpretation seems to me to underlie most textbook expositions of what is nowadays referred to as "the sampling theory approach" to testing hypotheses. It is possible to view this approach as an attempted synthesis, combining the evidential strands of Fisher's conception of significance tests with the more operational thrust of Neyman and Pearson. The contribution of the latter most often cited, however, is not their operationalism, but their provision of criteria for selecting the best test of given size, and, more particularly, their emphasis on the need to consider alternative hypotheses in gauging the evidence against a null hypothesis.[5]

Implicit in this oft-repeated tribute to Neyman and Pearson is the charge that Fisher's theory failed to provide workable criteria of optimality for a significance test and also failed to notice the importance of alternative hypotheses in gauging the evidence against null. That criticism, as I have already suggested, may well be wide of the mark, inasmuch as Fisher seems to have regarded the likelihood function as the appropriate index of *comparative* evidential support. Neyman and Pearson never really addressed the problem for whose solution Fisher proposed his theory of significance tests, namely, to assess the conformity between an hypothesis and data without reference to alternative hypotheses.[6] Be that as it may, we shall consider the eclectic approach on its own merits.

3. METHODOLOGICAL DIFFICULTIES

On what I am calling the 'eclectic interpretation', the import of an experiment for the considered hypotheses is dependent upon the error characteristics of the test. Now it is usually taken for granted in expositions of orthodox statistics that the test be specified in advance, and not only the test, of course, but the experiment, the sample space, the underlying probability model, the parameter space or set of admissible hypotheses, and even the 'tested' hypothesis. (It makes a difference to acceptance and rejection which of two considered hypotheses is labelled the 'test hypothesis'.) If one allowed the test employed to be chosen in the light of the observed outcome, one could not base a statement of evidence on the error characteristics of the post-selected test, presumably, because one's actual procedure is to select a test on the basis of what is observed. In the absence of any indication what, if any, systematic mode of selection is being employed, it is impossible to calculate the error characteristics associated with such a procedure

An alternative to this straight and narrow interpretation with its rigorous insistence on pre-designating all of the elements of the inferential process, would be to report exact significance levels. That is tantamount to regarding the exact significance level by itself as an index of the evidence against null, and there are grave and well-known objections to this. My own suspicion is that a reflective statistician who began by applying such an approach to testing in situations where well-defined alternative hypotheses with calculable likelihoods are present would inevitably be led to consider the likelihood function. In short, this path leads back to an essentially Fisherian position, or, at any rate, to a position which views likelihood as pivotal in comparative inference.

Appealing as consideration of the likelihood function may be, relatively few non-Bayesian statisticians have expressly adopted this stance.[7] Nevertheless, its influence is reflected in much current orthodox practice, for as we will see, it is difficult indeed to live within the strait-laced confines of a predesignationist methodology. The faintly derogatory overtones of 'eclectic' — 'opportunistic', perhaps even 'incoherent' — seem apt in conveying the flavor of current orthodox practice. Let us consider the pressures that militate against a rigid predesignationism.

A. Optional Stopping

Suppose that a decision maker believes the number of defectives present in a consignment to be small. He determines 1% as the cut-off or break-even point such that it pays him to accept the shipment when the lot proportion defective falls below this value and to return it otherwise, his losses being quite substantial if he accepts a consignment in which the proportion of defectives exceeds 1%. He is led on this basis to an acceptance sampling scheme which draws a large sample from each shipment and then accepts a shipment iff the sample proportion defective falls below a cutoff C (presumably less than 0.01). Upon sampling, however, he finds, contrary to expectation, that the proportion of defectives in this particular shipment is running much higher than 1%.

He would like to stop sampling, for he already has sufficiently definitive information to reach a decision, and further sampling would incur wasteful expenditure of time and money. The straight and narrow version of orthodox theory, however, forbids this; so-called 'optional stopping' of the sort here in question is expressly disallowed.

The dilemma can be made more piquant. The agent in question might have chosen a sequential sampling scheme, one that would allow him to terminate the procedure as soon as the sample proportion defectives exceeds a critical value B or falls below a second critical value A, leading to rejection of the shipment in the former case and acceptance in the latter. But imagine that our experimenter did not select a sequential plan. He cannot now pretend that he did — that would be cheating! To take seriously the selection of a comprehensive guide to subsequent action is to be *bound* by whatever strategy one selects. Even though our example is from decision theory, the same constraints would presumably operate in the inferential sphere, and for the reasons I cited above.

Consider another case of this sort. Experimenter 1, who is interested in the lot proportion defective, decides to sample 100 items and finds 2 defectives: experimenter 2, on the other hand, elects to sample until he finds 2 defectives, but, by coincidence, he happens to find his second defective upon examining the one hundredth item. The likelihood functions for the two experiments are

$$L(p) = \binom{100}{2} p^2 (1-p)^{100-2} ,$$

and

$$L(p) = \binom{99}{1} p^2 (1-p)^{100-2} ,$$

and these have the same kernal (viz., the part that depends on p). Hence, if both experimenters have the same prior distribution of p, their posterior distributions will also be the same. Thus, on a Bayesian analysis, the two men should reach the same conclusions about p. The principle here invoked is called the *likelihood principle*.[8]

From an orthodox standpoint, by contrast, different stopping rules were employed, and so the Bayesian equivalence of the two cases does not follow. In fact, it is easy to find cases where optimum orthodox tests for two experiments, one binomial, the other negative binomial, and a comparison of simple hypotheses, would disagree, one accepting the null hypothesis and the other rejecting it, despite the equivalence of the likelihood functions.[9] Orthodox theory thereby makes the import of the observed outcome depend on the entire sample space (through the critical region), and hence on outcomes that *might* have been observed. In the present instance, different *intentions* when to stop sampling are allowed to color the import of what is observed. Savage and other Bayesians have been quick to exploit this opening to contend that "the Bayesian approach is more objectivistic than the frequentist approach in that it imposes a greater order on the subjective element of the deciding person" (Savage, 1961, Section 5). But mightn't the conclusion be overdrawn? Consider the following point raised by Peter

Armitage in discussion with Savage (Savage, 1962, p. 72):

I think it is quite clear that likelihood ratios, and therefore posterior probabilities, do not depend on a stopping rule. Professor Savage, Dr Cox and Mr Lindley take this necessarily as a point in favor of the use of Bayesian methods. My own feeling goes the other way. I feel that if a man deliberately stopped an investigation when he had departed sufficiently far from his particular hypothesis, then 'Thou shalt be mislead if thou dost not know that'. If so, prior probability methods seem to appear in a less attractive light than frequency methods, where one can take into account the method of sampling.... Professor Savage ... remarked that, using conventional significance tests, if you go on long enough you can be sure of achieving any level of significance; does not the same sort of result happen with Bayesian methods? The departure of the mean by two standard errors corresponds to the ordinary five per cent level. It also corresponds to the null hypothesis being at the five per cent point of the posterior distribution. Does it not follow that by going on sufficiently long one can be sure of getting the null value arbitrarily far into the tail of the posterior distribution?

In replying, Savage invites us to consider an urn that either contains 3 red balls and a black, or 1 red and 3 blacks. In sampling with replacement, the experiment is thus binomial, with $p = 3/4$ or $p = 1/4$, and hence the likelihood ratio based on a sample of r red and b black balls reduces to 3^{r-b}. Suppose I wanted to sample until this ratio exceeded 10. Using common logarithms, this requires that $(r - b) \log 3 \geqslant 1$, or that $r - b \geqslant 3$. Even if I sample till doomsday, the probability of this happening when $p = 1/4$ is $1/27$ – my chances of deceiving you are minimal. But consider a significance test at some preassigned level, say 5%, and the most powerful 5% test of $p = 1/4$ at that. At a sample on n, there is already a 5% chance that my test will reject $p = 1/4$ when that hypothesis is true. But, in addition, if I am determined to sample until the test rejects $p = 1/4$, my test would have opportunities to mistakenly reject this hypothesis at smaller sample sizes on the way to n. So the actual probability of obtaining significance somewhere along the way in sampling n balls will be substantially higher than 5%. Whatever it is, it will increase if I sample two more balls, for if both new draws yield a red ball, some samples insignificant at sample size n will become significant at sample size $n + 2$. Hence, the probability of obtaining significance is cumulative, and, in fact, it tends to one. These results show why optional stopping must be disallowed. From the standpoint of significance testing, it really would be misleading to report a result as significant at the 1% level, say, and a sample of n, when in fact my intention was to sample either until a result significant at this level was obtained or until my sample reached $N \geqslant n$. The actual type I error probability associated with such a procedure would be considerably higher than the reported significance level, and, in extreme cases, might even approach one!

Now while this reply of Savage's is perfectly satisfactory as far as it goes, it is possible to carry these considerations a step farther. Imagine that I do decide to sample until the likelihood ratio in favor of $p = 3/4$ against $p = 1/4$ exceeds 10, and that, contrary to all reasonable expectation when $p = 1/4$, I succeed! I may, for example, obtain $r - b \geqslant 3$ for the first time after 10^{10} trials. Now it seems to me that some writers, possibly including Savage, go too far when they argue that this result, r red balls in 10^{10} trials, must be considered on its own merits as providing a likelihood ratio in favor of $p = 3/4$ in excess of 10. Given my intentions, I have actually excluded many possible binomial trial outcome sequences, so that, in effect, my sample space begins with the events *RRR, BRRRR, RBRRR, RRBRR, BBRRRRR*, etc. Now each of the events, qua ordinary Bernoulli sequence, is $3^{r-b} = 3^3$ as probable on $p = 3/4$ as on $p = 1/4$. Hence, in renormalizing so that the probabilities of these events (satisfying $r - b = 3$) conditional on each hypothesis sum to one, the normalization constants employed for $p = 3/4$ and $p = 1/4$ must stand in the inverse ratio of 3^{-3}. Hence, qua events of the new outcome space of sequences satisfying $r - b = 3$, each atomic event will have the same probability on $p = 3/4$ that it has on the hypothesis $p = 1/4$. Hence, whatever such event is actually observed, the likelihood ratio will be one! In short, the experiment thus hedged becomes totally uninformative. None of the elements of the reduced sample space agrees with $p = 1/4$, and so, following Popperian intuitions (which, however, were seen to be exceptionable in Chapter 6), we can say that, since no possible outcome will count as evidence against $p = 3/4$, the experiment cannot possibly confirm that hypothesis. It is irrelevant whether the likelihood ratio exceeds 10 after a small number of trials or after a huge number. This last result is slightly at odds with our feeling that a likelihood ratio of 10 after an enormous number of trials should rather lead us to suspect that the two hypotheses in question do not exhaust the possibilities, or that a peculiar sampling procedure has been employed, than that $p = 3/4$ is correct, for the expected likelihood ratio after a large sample would be well in excess of 10. This intuition is sound, but it applies only when we are conducting ordinary Bernoulli trials, not when there is a wholesale elimination of possible trial outcome sequences from the sample space. In any event, far from showing that the experimenter's intentions when to stop sampling are irrelevant to the import of what he observes, a more careful Bayesian analysis shows that, where those intentions materially reduce or alter the originally contemplated sample space, both the likelihood function and the import of the data are affected. We must in all cases inspect the actual likelihood functions before concluding that two experimental findings are or are not evidentially equivalent. I remark, in conclusion, that Savage's calculation of the odds of obtaining $r - b \geqslant 3$ when $p = 1/4$ also applies to an ordinary sequence of Bernoulli trials, with fixed probability $1/4$ or drawing a red on each trial. The calculation shows that if one went on sampling indefinitely, it is highly improbable that he would ever be led, by a likelihood criterion, to conclude that p was $3/4$ when in fact it was $1/4$.

B. *Randomization as a Basis for Inference*

In testing a simple hypothesis H against a simple alternative K (so-called 'simple dichotomy'), the most powerful α level test is constructed by including, among all those sample points not already included, those which maximize the likelihood ratio in favor of K. For example, the experiment comprises n Bernoulli trials, the sample space is $\{r: 0 \leqslant r \leqslant n\}$, and we wish to test H: $p = 0.5$ against K: $p = 0.7$. The critical region is constructed by first including $r = n$, then $r = n - 1$, etc., the process terminating when the prescribed type I error probability α has been attained. Thus, we seek r' such that

$$\alpha = \sum_{r=r'}^{n} \binom{n}{r} (0.5)^n ,$$

and the criterion counts values of r in excess of the cutoff r' as in poor agreement with H. The test may equivalently be described as a *likelihood ratio test*: the outcome is said to be in poor agreement with the test hypothesis just in case the likelihood ratio in favor of K is not less than the likelihood ratio in favor of K at the cutoff r'.

In practice, it often happens that there is no r' for which

$$\alpha = \sum_{r=r'}^{n} \binom{n}{r} (0.5)^n$$

holds exactly; at the last included point the size of the test, p', is just below the desired α, while if one more point were included the size, p'', would slightly exceed α. To obtain a test whose size is exactly α, Neyman and Pearson propose using a 'mixed' or 'randomized' test, which amounts to including a portion of that boundary point, as follows. Let t be the number between 0 and 1 for which $\alpha = (1 - t)p' + tp''$. Whenever the boundary point occurs a random experiment whose two outcomes have probabilities t and $1 - t$ is performed. If the former outcome occurs, the tested hypothesis is rejected, while if the latter occurs, it is accepted. It is easily verified that this test has size exactly α. The rationale here is that if one can tolerate a type I error probability up to 5%, say, then one should maximize power by using the mixed test, for leaving out the last point would give you more security against type I error than you require and less power than you could otherwise have.

As Fisher and others were quick to point out, the mixed test makes the import of the boundary point depend on an extraneous random experiment, like the flip of a coin. Mere consistency seems to demand that the same outcome observed by the same experimenter at different times should be accounted in agreement with the tested hypothesis on both or neither of those separate occasions. Yet, if error probabilities alone are to determine the import of what we observe, it is difficult to escape the conclusion that the boundary outcome does vary in its import with the flip of a coin. Another viewpoint can be defended, however, one that sees nothing sacrosanct about any particular significance level. Why insist on a test whose significance level is exactly α? Surely, in scientific contexts the tolerable type I risk cannot be ascertained with exactitude? On the other hand, one might ask, why insist on classifying each and every outcome as in agreement or disagreement with the tested hypothesis? Why not allow for equivocal or indecisive results? Only the mathematical difficulties, perhaps, prevented Neyman and Pearson from following this course. I will return to this matter of fixing or prescribing significance levels in Section 4.

Randomization, as in random sampling or random assignment of subjects to different treatments to be compared, plays another sort of role in statistics. Not all of the variables that can conceivably influence a variable of study can be controlled by the experimenter, and some can't even be detected. Randomization is designed to minimize the probability that such hidden variables mask the causal relationship of study. For example we might be interested in comparing the efficacy of two or more fertilizers on crop yield. We would select different plots to be fertilized by the different fertilizers that were as homogeneous as we could make them with respect to known or suspected sources of variation in crop yield — e.g. sunlight, rainfall, chemical composition prior to fertilization, and so on. Having secured maximal feasible homogeneity, we should still assign fertilizers to plots at random to insure that uncontrolled sources of variation have small probability of affecting yields of plots receiving different fertilizers differentially.

Suppose the plots are subdivisions of a single field, and that plots on the left-right diagonal happen to be especially propitious. If that should be found out, common sense tells us that the experiment is vitiated, when, perchance, fertilizer *A* had been assigned to just those plots. The fact that randomization made it improbable that one fertilizer would be assigned to the most propitious plots is surely irrelevant, once it is discovered that is what happened. Yet, again, if the import of the experimental findings is wholly a function of the error characteristics, one cannot dismiss the evidence for the superiority of fertilizer *A* after all. The results speak against the hypothesis of no difference by the lights of a criterion which has favorable error probabilities in 'repeated sampling from the same population'. And that is the end of the matter. It would seem that our interest in long-run chances precludes any interest in recognizably different subsets of the reference set (see Section 5).

C. *One-Tailed vs Two-Tailed Tests*

Often an experimenter suspects that a population mean is non-null, but he has no basis for predicting whether it is positive or negative. It seems, then, apropos of his state of knowledge to test $\theta = 0$ against the two-sided alternative $\theta \neq 0$ (a 'two-tailed test'). He will often find, though, that the more powerful one-tailed test against a directional alternative suggested by his data would have rejected the null hypothesis where the weaker two-tailed test did not. Switching from a two-tailed to a one-tailed test, after observing the results, would, however, clearly violate the prescription that the test be selected prior to experimentation. The actual procedure employed would be noting the direction of departure from null and then testing the null hypothesis against the indicated alternative át the prescribed significance level. Marks (1951) alleges that such procedure inflates the actual probability of type I error, in fact, doubling its actual value:

> It must be emphasized that the one-tailed test is not justified unless the prediction is made prior to the data. If an investigator . . . on studying these results generates a theory which will account for them, he cannot accept such afterthoughts as predictions, i.e., switch – on the spot – from a two-tailed to a one-tailed test . . . . He can use the results to predict *future* data. If he makes the switch after collecting the data, he has, in effect, increased the probability of committing a Type I Error without realizing it. If he uses the 5% level on a one-tailed test, this is equivalent to the 10% level with a two-tailed test, and he has unwittingly doubled the level of confidence: a dangerous departure from scientific conservatism.

His argument seems to be that in following the 'switch' procedure, the experimenter is *really* using a two-tailed test. Marks seems to believe that if one predicts the direction of departure from null beforehand, one is willy nilly doing a one-tailed test (and he equates this with confirmatory experimentation); by the same token, without a prior prediction, one is willy nilly using a two-tailed test (and doing exploratory testing). In the case of normal populations Marks has uppermost in mind, $P(x > k\sigma/\mu = 0) = P(\|x\| > k\sigma/\mu = 0)/2$ and so the two-tailed test, which Marks alleges the 'switch' strategist is really employing, has twice the prescribed or desired type I error probability.

To calculate the type I error probability for the 'switch' procedure, I make the assumption that the observed departure from null is positive or negative with equal probability (a dispensable assumption). Then the experimenter does one or other of the one-tailed tests about equally often. But since both tests have type I error probability α, so must the 'switch' test (which is rather like a mixed test, except that the test to be performed is determined by the observed departure from null rather than by the flip of a coin). So the type I error probability is not doubled, as Marks alleges. At the same time, the 'switch' test most assuredly does run the additional risk of erroneously rejecting $\mu < 0$ when $\bar{x} > 0$, and *vice versa*. (Perhaps, in the present context, this should be labelled a 'type III error'.) Without a prior distribution of μ, nothing can be said about the magnitude of this error probability, beyond the fact that it is no larger than the indicated significance level at each value of the mean less than 0.

So there is something to the claim that the 'switch' test inflates risk (though not type I risk). On the other hand, the very same risk may well be present when a prior prediction is made. Can we really be sure that $\mu < 0$ is impossible? If there are theoretical grounds for excluding it, they might well occur to the experimenter after looking at the data. In that case, his actual risk should not be any larger than that of the man who 'makes a prediction'. Perhaps the real difference at issue is that between a man who guesses that $\mu > 0$ *merely* on the basis of a small sample, and a man who has additional theoretical grounds upon which to base a prediction. But then, as J. M. Keynes pointed out long ago, the point has nothing to do with predesignation, but with whether or not theoretical grounds or corroborating empirical evidence can be adduced. The conclusion should then be that you can switch, but only when such grounds can be given for believing one of the directional alternatives. On a Bayesian analysis, this amusing conundrum never arises: one simply obtains a posterior distribution of the mean from which the probabilities of the three alternatives $\mu < 0$, $\mu = 0$, $\mu > 0$ can be calculated. I expressly allow the prior of μ to be modified in the light of new theoretical arguments, as well as updating it by conditionalizing on new data.

In the cases most frequently encountered, a lump of prior probability mass will be concentrated at the null value.

D. *Violation of Assumptions*

I said earlier that it is difficult to live within the confines of a predesignationist methodology. Actual orthodox practice fully bears this out; indeed, standard orthodox texts are all replete with post-designated tests, many of which seem inescapable. In analysis of variance, for example, upon rejecting the hypothesis that all means are equal, orthodox texts show you how to go on to test other more particular hypotheses about the means suggested by the data, and in particular, to decide which of the means is largest. (How is this any different from using the observed direction of departure from null to select a one-tailed test?) Those same texts — and their number is legion — also show how to test the underlying assumptions of the usual analysis of variance model, viz., normality and equality of the variances (homoscedasticity), again using the same data. Similarly, they show how to test underlying assumptions of randomness, independence and stationarity, where none of these was the predesignated object of the test (the 'tested hypothesis'). And yet, astoundingly in the face of all this, orthodox statisticians are one in their condemnation of 'shopping for significance', picking out significant correlations in data *post hoc*, or 'hunting for trends in a table of random digits'. With regard to the last example, we are told that a finite section of any such table is bound to manifest some pseudo regularities which disappear when larger sections of the same table are examined. Thus, they are of no account unless they were predesignated and made the 'tested hypothesis' prior to collection of the data or examination of the sample. An especially bizarre illustration of this kind of thinking may be found in the reactions of professional statisticians to the U.S. draft lottery of some years back. Capsules containing January dates were placed first in a large urn and shuffled, then capsules containing February dates, and so on up to capsules containing December dates placed last. If shuffling were inadequate — as it often is — one would expect capsules placed first to sink to the bottom and be drawn last. That is precisely what occurred (cf. Example 2 of Chapter 9). Yet, the comments I read afterwards all reflected the conventional wisdom that any such lottery was bound to manifest pseudo regularities of this sort, and, not having been predesignated, the observed such regularity was no evidence that the lottery was not truly random. (One hard-pressed official defended it by contending that one's birthday is already the outcome of a random lottery anyway!)

Imagine that we were testing an hypothesis about a binomial parameter or a normal mean, but afterwards, upon examination of the data, it appeared that the trials were not truly random or the population not truly normal. Again, these latter assumptions, not being themselves the object of the test, cannot be said to be counterindicated by the present data. Instead, a new experiment must be designed to test whether they are realistic assumptions about our current experiment! Running another experiment might, of course, turn out to be prohibitively expensive or time-consuming. But, practical difficulties aside, such a procedure relies on the tacit assumption that we can replicate present experimental conditions nearly enough for that subsequent experiment to shed real light on whether our assumptions hold in the present experiment. It is little wonder that Orthodox texts tend to be highly ambivalent on the matter of predesignation.

4. SIGNIFICANCE LEVEL AND POWER

The eclectic interpretation perceives strong evidence against the null hypothesis in an outcome that falls in the critical region of a test with 'good' error characteristics. On the face of it, a test of small size and high power has good characteristics, but there is more to this question than meets the eye.

Neyman (1950) offers the following recommendations. First, in a dichotomous comparison, call that hypothesis whose mistaken rejection would incur greater loss or disutility the *tested hypothesis*. (E.g., in a medical diagnosis, a false negative — classifying an infected patient as healthy — would likely have more dire consequences than a false positive — classifying a healthy patient as infected.) Having labelled the tested hypothesis, fix an acceptable level of type I error probability — the size of the test — and among all tests of that size (including randomized tests), choose that with highest power.

Now one cannot reduce size and increase power simultaneously without also increasing the sample size of the experiment. The consequence of fixing a tolerable level of type I error probability, at a fixed sample size, might be to drive the type II error probability unacceptably high. It is therefore implicit in Neyman's recommended procedure that one simultaneously adjust three quantities: the two error probabilities (or size and power) and sample size. In Bayesian decision theory, an optimal experiment can be determined, taking due account of sampling costs (Chapter 1). But no guidelines exist for the analogous problem of choosing a triple (α, β, n) in Neyman-Pearson theory, where α and β are, resp., the type I and type II error probabilities, and n the sample size.

These judgements are left to the individual experimenter, and they depend in turn on his personal utilities or rates of exchange between the different costs and benefits. But if we are 'interested in what the data have to tell us', why should one experimenter's personal values and evaluations enter in at all? How, in short, can we provide for objective scientific reporting within this framework? Reading passages like the following (Neyman, 1957, p. 18), one realizes that a vital distinction, that between the private or personal uses of statistics for individual decision making or sharpening one's own hunches and the public function of data analysis in conveying 'what the data have to tell us', is lost when one bases all of statistics on 'inductive behavior':

Whatever the scientist's beliefs and preferences may be, the knowledge of the performance characteristics of all possible decision rules will allow him to choose the one that fits his case best.

In practice, the significance level at which a tested hypothesis is rejected is interpreted, willy nilly, as indicating the strength of the evidence against the hypothesis.[10] Indeed, controversy has arisen in the social sciences over what appears to have been a deliberate publication policy on the part of some journals to exclude findings that were not significant at a minimally stringent level of 5%.[11] Such automatism belies Neyman's envisaged tailoring of significance tests to the circumstances, preferences and beliefs of the individual scientist, but it is perhaps inevitable in the absence of a satisfactory basis for assessing the evidential import and scientific significance of statistically significant findings.

The problem of interpretation is no mere academic quibble. In the first place, the meaning of a significant finding depends as much on the sample size as it does on the indicated (or preselected) significance level.[12] Secondly, in a variety of identification and classification problems, the significance level employed gives no clue as to the proportion of correct answers the test will yield. Finally, prior knowledge can (and should) influence both the choice and interpretation of the test. As C. A. B. Smith noted (Smith, 1959, p. 292):

A significance test . . . may 'reject a hypothesis at a significance level P', but P here is not the probability that the hypothesis is true, and indeed, the rejected hypothesis may still be probably true if the odds are sufficiently in its favor at the start. For example, in human genetics, there are odds of the order of 22 : 1 in favor of two genes chosen at random being on different chromosomes; so even if a test indicates departure from independent segregation at the 5 per cent level of significance, this is not very strong evidence in favor of linkage.

Smith's point is that, given the small proportion of linked pairs of genes, the bulk of the pairs so classified by a test with a type I error rate even as high as 5% will be spurious linkages. This point is of capital importance; I return to it below.

My second point is amply illustrated in the next chapter with reference to the problems of identifying the degree of a polynomial regression and the order of a Markov chain. A much simpler example will serve to make this point, however, namely, the problem of testing a simple hypothesis against a composite alternative. Given a binomial distribution, for example, we wish to test $p = 0.5$ vs $p \neq 0.5$. Let us suppose that a miss is as good as a mile: the scientist is interested only in maximizing his proportion of 'hits' (correct identifications). To give the problem a more concrete form, imagine that we are given a population of urns containing black and white balls in various proportions, where p is the proportion of black balls in a chosen urn. We are given that half the urns have $p = 1/2$, while all other possible values of p occur with equal frequency among the remaining urns. A small sample is drawn from the chosen (at random) urn, and it is classified as a $p = 1/2$ or a $p \neq 1/2$ urn. A dollar is won or lost according as the urn is correctly or incorrectly classified. It is easily seen[13] that expected income is maximized by using the Bayes test which classifies the urn as a $p = 1/2$ urn or not according as the ratio of the average likelihoods of these two hypotheses, viz.,

$$(1/2)^n : \int_0^1 p^x (1-p)^{n-x} \, dp$$

exceeds unity or not, where x is the number of black balls drawn in the sample of n. Hence, the urn is classified as a $p \neq 1/2$ iff

$$(10.1) \qquad \binom{n}{x} < 2^n/(n+1),$$

i.e., $p \neq 1/2$ is favored just in case the observed binomial coefficient, $\binom{n}{x}$, is smaller than the average of the binomial coefficients. The significance level of an UMP unbiased test, even combined with the sample size, gives no clue to the proportion of 'hits'; I would have no idea how to choose a significance level for this problem (save, via the Bayes test).

The previous point is related to the first point above. By increasing sample size, we make a test of a sharp null hypothesis ever more sensitive to small (and perhaps practically or scientifically unimportant) departures from null, whence the 'paradox' that obtaining significance with a small sample is better evidence against the null hypothesis than significance with a large sample.[14] Hodges and Lehmann (1954) attempt to surmount this difficulty by testing an approximate null hypothesis. But there are two strong objections to their proposal. While it is true that an approximate null hypothesis is more sharply discriminable from its alternative than a sharp null hypothesis, still, science abounds in cases (e.g., the familiar Mendelian ratios) where simple values are indicated by theory. In genetic linkage, for example, the hypotheses $\theta = 1/2$ and $\theta \neq 1/2$ even point to different physical states: two genes lying on the same or on different chromosomes.

My strongest objection to the Hodges and Lehmann proposal, however, is that it doesn't address what I take to be the actual difficulty here. It is less that a test based on a large sample will detect unimportant departures from null when the null hypothesis is approximately true than that it will find in favor of a (simple or composite) alternative hypothesis when the evidence overwhelmingly *favors* the null hypothesis.

This is even true when we are testing one sharp hypothesis, say, $H: p = 0.5$, against another, say $K : p = p_1 > 0.5$. The boundary point of the critical region of the most powerful 5% test is close to $0.5n + 1.64(n \times 0.5 \times 0.5)^{1/2} = 0.5n + 0.82n^{1/2}$, using the normal approximation, n the sample size. At this point the support (log likelihood ratio) for K against H is

$$\log \frac{p_1^{0.5n+0.82n^{1/2}}(1 - p_1)^{0.5n-0.82n^{1/2}}}{0.5^n} = An + Bn^{1/2}$$

after collecting terms, with $A = \frac{1}{2}[\log p_1 + \log(1 - p_1)] - \log\frac{1}{2}$. Using the concavity of the log function, $A < 0$, and since a polynomial in n is, for sufficiently large n, monotonic in the sign of its leading coefficient, $An + Bn^{1/2} \rightarrow -\infty$ as $n \rightarrow \infty$. Hence, the support for H tends to infinity with the sample size at the boundary point of the critical region. The same will be true, of course, in testing H against a composite K. Even points rather removed from the boundary can favor the null hypothesis. Thus, at $n = 1000$, the boundary point of the most powerful 5% test of $p = 0.5$ vs $p = 0.7$ is $x = 526$. Yet, the likelihood ratio in favor of $p = 0.5$ is 2.89×10^{19} even when $x = 550$.

The criticism is not purely external. For, as Spielman (1973) points out, the reliability of a most powerful test at points on or near the boundary is well below the advertised type I error probability. To see this, note that the ratio $P(x^*, H): P(x^*)$ gives the percentage of times H is true when the boundary point x^* is observed, and hence the percentage of erroneous rejections of H when $x = x^*$. When the considered hypotheses, H, K, are equiprobable, this ratio reduces to $P(x^*/H)/[P(x^*/H) + P(x^*/K)]$, which is close to one, since $P(x^*/H)$ is much larger than $P(x^*/K)$, at large n. Even at $n = 100$, $x^* = 58$ (in our example), and the ratio in question is 0.87, which means that $p = 0.5$ would be erroneously rejected about 87% of the time. True, this figure assumes that H and K are equiprobable, but other possible prior probabilities give similar results. Even if $P(H) = 0.1$, so that x^* is more often observed when K holds, H would be erroneously rejected about 44% of the time at x^*. Unreliability at the boundary is entirely compatible with good error characteristics (and, in particular, with low type I error probability), for boundary outcomes rarely occur. At extreme outcomes, which also occur infrequently, the frequency with which the test commits type I errors will be far less than the advertised type I error rate; the extreme outcomes counterbalance the boundary outcomes so that the overall frequency of type I errors does equal the advertised rate. Now as Hacking (1965) argues, unreliability at the boundary is of no importance when the test is used as a routine signaling device, as in industrial sampling inspection, where our interest lies in achieving satisfactory quality control over the long haul. But inferential problems are not like that. Having been misled by a test at a boundary point, it is little comfort to be told that such outcomes will occur infrequently in a long sequence of imaginary repetitions of the experiment that will never be performed. There is a 'recognizable subset' of cases for which the probability of being misled by our test is different (cf. Section 5).

Unreliability at the boundary results at large samples, however, only if the significance level is held fixed,[15] and this naturally suggests reducing the significance level as the sample size is increased. This suggestion is put forward by Kendall and Stuart (1967), p. 183. They write:

> ... we can determine α in any way we please, and it is rational ... to apply the gain in sensitivity arising from increased sample size to the reduction of α as well as of β. It is only the habit of fixing α at certain conventional levels which leads to the paradox.

But Kendall and Stuart say nothing about the rate at which α should be lowered, and so their recommendation adds still another factor determining the import of the data to be adjusted at the discretion of the individual investigator. It is particularly striking, in this connection, that the Bayes test, (10.1), reduces the significance level automatically with increasing n, and does so *in a mathematically determinate way*. Thus, the Bayes test has size 10% at $n = 10$, and size 5% at $n = 100$, etc., always coinciding with the UMP unbiased test of that size.

I want now to take up two more proposals contained in Morton (1955) and Giere (1976). Giere draws on the familiar distinction between exploratory and confirmatory testing. Where no workable theory is at hand, we simply want some results to think about. At this exploratory stage, he says, stringent tests are not desired, but high power is: we want to be sure of detecting whatever first order effects are present. Thus, a satisfactory test will have low power, say less than 0.05, against materially insignificant departures from null and high power, say at least 90%, against materially significant deviations.

Just this strategy is proposed and illustrated in Morton (1955), a paper devoted to sequential likelihood ratio tests for human linkage. It is well known that sequential tests reduce the average sample size (ASN) as compared to fixed sample size tests with like error characteristics (Figure 4, Morton, 1955), but even at that, the ASN needed to satisfy the requirements of high power against materially significant departures from null can turn out prohibitively large. Thus, in the linkage problem, good power against the value $\theta = 0.4$ of the recombination fraction requires an ASN of 5700 (and a fixed sample size of roughly twice that number). It is by no means obvious, even with knowledge of the sampling costs, what feasible combination of size, power and ASN to choose, and even less obvious that one man's personal trade-offs should determine the status of an effect (e.g., a putative linkage) as accepted or rejected.

In the face of this, Giere blandly claims that the required adjustments are controlled by 'the aims of inquiry', that "they are the kinds of judgements on which most experienced investigators should agree" (p. 79). I find this claim faintly incredible, and to illustrate my doubts, let me juxtapose two passages from the very articles under consideration.

Giere (p. 78) writes:

At the beginning of a program of inquiry, when very little is yet known, one can afford to sacrifice some precision and reliability simply to have some results to think about. A partly misleading basis for theoretical speculation is better than no basis at all.

Morton (p. 284) writes:

We are especially anxious to avoid the assertion that two genes are linked when in fact they are not, since a misleading linkage map is worse than no linkage map at all.

Giere insists that "one really does accept statistical hypotheses (and the underlying causal hypotheses) as being true for the purposes of inquiry", and 'accepted' hypotheses are to be "regarded as something that *must be explained* by a proposed general theory of the relevant kind of system . . . and assumed true when we design new experiments to test further hypotheses" (p. 79; my italics). Literally adhered to, Giere's recommendation would amount to a commitment on the part of a group of researchers (e.g., a subdivision of the American Psychological Association) to explain or premiss any apparent effect or causal relation detected by a test whose error characteristics and sampling costs were collectively deemed satisfactory.

Scientists simply do not make such commitments (nor should they). Evidenced effects are no doubt *prima facie* explananda, but if a theory cannot account for every putative effect reported in the literature, that is surely no bar to its being taken seriously. The Copernican theory was unable to account for the observed variation in the brightness of Venus, among other things, but that apparent failure did not rule it out of court. Rather, a theory which fails to fit a prima facie finding incurs what we may call an 'evidential debit' – the stronger the evidence, the greater the debit. The debits and credits a theory accumulates over time are sufficient to condition appropriate judgements of its promise for future research. The proposals of Giere, Popper, Lakatos, and others to lay down acceptance or rejection rules in advance would, if followed, hamstring science. This is concealed in all cases by some strategic hedging or waffling, as when Giere assures us that mistaken acceptances "need not be irreparable" (p. 79). But one can't have it both ways. If accepted hypotheses are presupposed in subsequent inquiry, there will be no occasion to test them further. On the other hand, if their dissonance with a new and otherwise promising theory is sufficient to call them into question and cause them to be tested anew, what then becomes of Giere's contention that hypotheses are 'really accepted' and not merely evidenced or confirmed? The crying need, it seems to me, is not to enforce consensus via pre-selected acceptance rules, but to objectively measure the evidential debits a theory incurs. I proposed observed sample coverage to this end in Chapter 9.

The Bayesian way with linkage is to compute an ongoing posterior distribution of the recombination fraction θ. Since there are 22 autosomal chromosomes in a haploid set, the prior will assign mass $21/22$ to the value

$\theta = 0.5$, corresponding to no linkage. It seems reasonable to distribute the remaining $1/22$ uniformly over the values of θ less than 0.5. Morton (pp. 281ff.) offers a detailed derivation of this prior, and since it is based on empirical data, it should be used in place of an uninformative prior for this problem. Data from different experiments are easily combined, and this allows one to keep track of the odds favoring different putative linkages over time.

Morton, however, argues for the use of sequential tests, claiming their superiority to Bayesian methods. This is puzzling, for Bayesian methods are *ipso facto* sequential: even optional stopping is allowed. It is rare indeed to find a writer who uses a prior without using Bayes' rule. The particular use Morton makes of the prior is of interest, however, for it is responsive to the point of C. A. B. Smith's alluded to above.

Smith, you recall, notes that if the a priori chance of linkage is 1 in 22 (for autosomal traits in humans), a test with a 5% type I error rate will 'detect' more spurious linkages than genuine ones. Giere's proposal is insensitive to the a priori chances; their palpable relevance further belies his contention that 'the goals of inquiry' are a sufficient basis upon which to adjust significance level, power and sample size. Morton, by contrast, proposes to take cognizance of the a priori chances by calculating what he calls (p. 284) "the posterior probability of type I error". Namely, let α be the type I error probability, and ϕ the a priori chance that two genes are linked. Then the proportion ρ of spurious linkages among those pairs of traits so classified is given by:

$$(10.2) \qquad \rho = \frac{P(\text{Classified linked/not linked})P(\text{not linked})}{P(\text{classified linked})} = \frac{\alpha(1-\phi)}{\alpha(1-\phi) + \phi\bar{P}}$$

where $\bar{P}$ is the average power of the test (the probability of detecting linkage when it is present). One might call $1 - \rho$ the *reliability* of the test, and $\bar{P}$ its *sensitivity*. Clearly, it is the reliability of the test (the percentage of genuine linkages among those 'detected') that matters, and not the type I error rate as such. (This further instances my earlier remark that the error probabilities seldom tell you what you really want to know about the prospective performance of a test.) Similarly, in screening for a disease, it is not the probabilities of false positives and false negatives that matter, but the percentage of infected persons among those showing a positive reaction, viz.:

$$P(\text{infected/positive}) = \frac{P(\text{positive/infected})P(\text{infected})}{P(\text{positive})},$$

which clearly depends on the proportion of the population infected (which can often be estimated with considerable accuracy).

Now the burning question that confronts Morton is: what has been gained, from a non-Bayesian standpoint, if the calculation of reliability is based squarely on a prior distribution? In screening a population for a disease, reliability, sensitivity and sample size determine a reasonably well-defined decision problem. In human linkage studies, however, we are only interested in the strength of the evidence favoring various putative linkages, and, presumably, one cannot make better use of the a priori chance of linkage than to apply Bayes' rule, updating the odds with each new addition to the data. Only the presumed necessity of reaching decisions, I think, leads Morton to conclude otherwise.

Let me end this section with some comments on Giere's discussion of confirmatory testing. He views "the process of taking a clearly formulated theory and deducing from it (with appropriate auxiliary theories) a statistical hypothesis" as "the operation of a binomial system . . . having an unknown binomial propensity for producing true statistical hypotheses" (p. 83). Suppose, for illustration, that this propensity is 50%. Then, as Giere notes (p. 84), if the power of a test against clearly divergent alternatives is at least 90%, our chance of detecting the falsehood of the theory in one trial is only 45%. By testing several independent consequences of the theory, however, power may be greatly augmented, provided we agree to reject the theory just in case *any* of its predictions is rejected. To be sure, the type I error probability is thereby increased, so that if we want a test, based on, say, five independent predictions, to have size at most 5%, each of the separate tests must have a much smaller size. Indeed, if the size of each of the five tests is 1%, and each has power 90%, the power of the composite test is

$$1 - 0.5^5 \left[\binom{5}{0} 0.99^5 + \binom{5}{1} 0.1(0.99^4) + \cdots + 0.1^5 \right] = 0.9519.$$

For the falsehood of the theory is *not* detected iff either all five predictions are correct and none are erroneously rejected, or one is incorrect but not detected as such, while the remaining correct predictions are not erroneously rejected, etc. Similarly, the composite test erroneously rejects the theory when at least one of the five (necessarily correct) predictions is erroneously rejected, hence the size is $1 - 0.99^5 = 0.049$. This illustrates the increased sensitivity that results by using several independent tests. Yet, several features of this approach are disturbing.

Scientists seldom, if ever, pre-determine the number of predictions whose verification they will regard as establishing a theory. Now while I agree that "it only takes a few good successful predictions to establish a theory as the working paradigm for a scientific field", it is not clear that scientists can or should state how many (or which) in advance. Yet, if the point of testing many predictions is to improve the operating characteristics of the resulting composite test, the operational interpretation of these characteristics does require that the test (and, in particular, the number of tested predictions) be stated in advance. Giere writes (p. 86): "I do not mean to imply that scientists in any sense ever pick a fixed sample of predictions". But, again, he can't have it both ways.

The "unknown binomial propensity of a theory to yield true predictions" is also bothersome. Giere writes (p. 85): "The theory is either far from true or approximately true". But he offers no measure of distance from the truth, and is well aware how problematic this notion is. He says: "the best way I know of dealing with this difficult, and philosophically suspect, notion [nearness to the truth] is to think in terms of the propensity of the testing process to yield a clearly false statistical prediction" (p. 84). That does not so much deal with verisimilitude as sweep it under the rug. The propensity in question depends on: (i) the (problematic) proportion of false consequences of the theory, and (ii) the sampling rule. Both of these are anybody's guess. It is not even clear, in the light of the following passage (p. 85) that the propensity in question is truly binomial:

To obtain a sufficiently powerful test of the theory we should test new consequences, ones not previously known to be true, or at least ones not considered in designing the theory. Indeed, a good methodological rule is to test consequences that experienced investigators would judge likely to be false. Using such judgements, which need not be interpreted as expressing subjective degrees of belief, may increase the power of the test of the theory, *though by how much will be unknown* [my italics].

I have already commented sufficiently on the claim that theories are more strongly evidenced by verification of their 'novel' (as opposed to, improbable) predictions (Chapter 7). Here it is more in point to ask what use is to be made of a test whose operating characteristics are unknown (indeed, unknowable)? If we deliberately select consequences judged probably false, are we not altering the propensity of the testing process to issue in false predictions? What reason is there to believe that this propensity will be stationary from test to test? Again, while I am obviously in sympathy with the suggestion that probability judgements need not be subjective, the judgements Giere cites are another matter.

That the Neyman-Pearson theory may wholly fail to address the problem for which Fisher's theory of significance tests was designed has already been stated (Chapter 9). Where confirmatory testing is in question, we often lack any clearly defined alternative theory. Take so unprepossessing an example as Benford's law (Chapter 3, Example 6). This law states, you recall, that k appears as first digit in a tabular entry with probability $p_k = \log(k + 1) - \log k, k = 1, \ldots, 9$. What are the alternative hypotheses against which we desire high power? One could doubtless formulate any number of alternatives, but that is just the trouble: there are no *salient* (i.e., simple and plausible) alternatives that suggest themselves. Considerations of power seem otiose in such a context, and, in fact, most statisticians would be content with a chi square test. There is in this case at least a fairly well-defined set of predictions. But intuitions about which scale-invariant tables are most likely to violate the law, much less what proportion of them will, are fuzzy in the extreme. Far better, I think, to apply the methods of Chapter 9: to compute the observed sample coverage and thereby the improbability of the theory's accuracy. 'Theories' run the gamut from empirical laws (Benford's, Kepler's, etc.) to simple models of phenomena (e.g., the two-factor model of the $A-B-O$ blood groups, the Markov model of concept identification) to more abstract constructions (Relativity, Quantum theory) to theoretical blueprints or research programs (the atomic theory, the particulate theory of inheritance). The absence of well-defined alternatives appears at every level on this scale of abstraction; confirmatory testing is not always comparative testing. And where well-defined alternatives have not yet crystallized out, the outcome probabilities Giere requires become tenuous indeed. In any event, it is by averaging the likelihood function against a prior distribution of the considered theories that these probabilities are computed. But if a prior is presupposed, the question again arises what has been gained from a non-Bayesian standpoint?

5. THE REFERENCE CLASS

Frequency theories of probability have always confronted the twin problems of the reference class and the single case. These difficulties, as we have already seen, beset the sampling theory approach to testing hypotheses at every turn. The difficulties are accentuated in the context of testing hypotheses by the essential arbitrariness of the reference set in which the experiment is embedded.

We want to say, roughly, that if an item is chosen at random from a reference class p per cent of whose members have property Q, then that individual item (or single case) has Q with probability p. To borrow an example from Fisher (1958), male births are known to occur slightly in excess of 51% of the time, so our principle leads us to conclude that Mary's first child, shortly to be born, will be male with probability about 0.51. Experience might also show, however, that first births have a higher male: female ratio than births in general. In Fisher's words, we may be able to pick out a 'relevant subset' of the reference set containing the single case in question and having a higher or lower proportion of Q's. We can apply the single case rule, then, only when no such relevant subset can be recognized. Skeptics may doubt whether this condition is ever met, but I think they may be confusing the existence of relevant subsets with their recognizability. The point I would urge, rather, is that the qualifier Fisher appends here is but a disguised form of the principle of indifference (though, unsurprisingly, Fisher vehemently denied this). We are arguing from ignorance of any relevant subset to a given probability attribution. (Fisher himself habitually referred to this as 'a postulate of ignorance'.)

Be that as it may, Fisher's proviso of non-recognizability is often violated in orthodox (Neyman-Pearson) testing. As we saw in the previous section, an optimum Neyman-Pearson test is appreciably less reliable on the recognizable subset of boundary points of the critical region. It is also violated by the so-called *compounding* device due to Herbert Robbins (Robbins, 1963). By combining tests for totally unrelated experiments, one can obtain better error characteristics than are obtainable for those experiments treated individually. Neyman himself hailed this contribution of Robbins' as a 'major breakthrough'; other statisticians have hailed it as the *reductio ad absurdum* of the 'behavioral' approach. It was in this spirit that Cox (1958) offered the following example.

We are given two normal populations with the same mean θ and different variances, with $\sigma_1^2 \gg \sigma_2^2$. A fair coin is flipped, and we sample one population or the other, depending on whether it lands heads or tails. We wish to test the hypothesis $\theta = 0$ against the simple alternative $\theta = \theta' \doteq \sigma_1$, at, say the 5% level. It is assumed that we know which of the two populations it is we are sampling.

A conditional test bases calculations of size and power on the particular population known to have been sampled, and thus rejects the null hypothesis if $x > 1.64\sigma_1$ when the first population (θ, σ_1) is sampled, and rejects it if $x > 1.64\sigma_2$ when the second population (θ, σ_2) is sampled. However, the critical region defined by

$$x > 1.28\sigma_1, \text{ if the first population is sampled,}$$

$$x > 5\sigma_2, \text{ if the second population is sampled,}$$

is easily seen to have the same significance level (roughly), since the type I error probability goes up to about 10% when (θ, σ_1) is sampled but goes down to virtually 0 when (θ, σ_2) is sampled. On the other hand, this mixed (or unconditional) test is seen to have higher power, since $\sigma_1 \gg \sigma_2$. About this example, Cox writes:

... if the object of the analysis is to make statements by a rule with certain specified long-run properties, the unconditional test just given is in order, although it may be doubted whether the specification of desired properties is in this case very sensible. If, however, our object is to say 'what we can learn from the data we have', the unconditional test is surely no good. Suppose that we know we have an observation from (θ, σ_1). The unconditional test says that we can assign this a higher level of significance than we ordinarily do, because if we were to repeat the experiment, we might sample some quite different distribution. But this fact seems irrelevant to the interpretation of an observation which we know came from a distribution with variance σ_1^2 ....

To sum up, if we are to use statistical inferences of the conventional type, the sample space must not be determined solely by considerations of power, or by what would happen if the experiment were repeated indefinitely. If difficulties of the sort just explained are to be avoided, Σ *should be taken to consist, so far as is possible, of observations similar to the observed set S, in all respects which do not give a basis for discriminating between the possible values of the unknown parameter θ of interest* (my italics). Thus, in our example, information as to whether it was (θ, σ_1) or (θ, σ_2) that we sampled tells us nothing about θ, and hence we make our inference conditionally on (θ, σ_1) or (θ, σ_2).

As Cox also notes, the underscored principle was clearly formulated by R. A. Fisher in several places (e.g., Fisher (1959), Chapter 3, or Fisher (1955)), and, moreover, Fisher offered many practical illustrations of it by way of criticism of the Neyman-Pearson approach. More recently, Giere (1976), while granting the intuitive force of examples like Cox's, has sought another basis for preferring the conditional to the unconditional test — a basis that does not invoke Fisher's *ancillarity principle*, as it has come to be called. The latter, as Birnbaum (1962) showed, is entailed by the likelihood principle, and so follows easily on a Bayesian analysis (as should be intuitively obvious). The point of Giere's exercise, then, is to dilute the force of Birnbaum's observation by showing how to handle Cox's example within an orthodox setting, and more specifically, by appealing to a 'propensity interpretation' of probability tailored to apply to the single case. I quote

Giere's analysis of Cox's example below:

... let us follow in detail the course of a single trial .... First is a turn on the randomizer which has equal propensities for each of two possible outcomes. One outcome is linked with the activation of process one; the other with process two .... Assume we desire the best 5% test of the hypothesis $\mu = 0$ against $\mu = \mu' \doteq \sigma_1$. Given that we *know* which process has been activated, we know whether the test process refers to the propensity distribution (μ, σ_1) or (μ, σ_2). Thus we know which process was causally operative in producing the data, and consequently, the decision to accept or reject H. Equally we know which process was *not* operative. This provides the needed basis for declaring that features of the non-operative system must be irrelevant to our conclusion concerning the mean value of the chosen process on this particular trial. That other system was not a causal part of this trial.

The point seems to be that distinct physical processes are involved, and this allows one to say that different 'propensities' are in question. Finally, the significance test we use is determined by the operative propensity. A similar analysis applies to a simpler example discussed by Basu and Birnbaum, where a randomizer selects whether we observe 10 or 100 Bernoulli trials. There, too, a mixed test of given size has higher power, but there, too, the 10-trial process has a different 'propensity distribution' than the 100 trial process.

Well might one wonder what is the point of dragging in these so-called 'propensities'. Couldn't a frequentist resolve the difficulty by 'recognizing a relevant subset'? No doubt he could, but then he could also recognize the relevant subset of a critical region comprising the boundary points. So he could accommodate everyone's intuitions about Cox's example, but only by throwing out error probabilities altogether as a basis for inference. It is because there are two physically distinguishable processes before us in the examples of Cox, Basu and Birnbaum, but only one in the case of tossing a coin (and rejecting a null hypothesis at boundary points of the critical region) that the propensity theory is apparently able to save the day.

Unfortunately, this triumph of the propensity interpretation is illusory. The illusion persists only for as long as we confine attention to illustrations of Fisher's ancillarity (or conditionality) principle which happen to involve physically distinguishable populations or processes. *Most* of the illustrations Fisher and others have given do not. One such is the notorious problem of testing the equality of two normal means when the variances (and the ratio of the variances) is unknown. The test Behrens and Fisher proposed is conditional on the observed ratio s_1/s_2, where s_1^2, s_2^2 are the sample variances. Here, too, there is a more powerful unconditional test, due to Welch, that, for a time, was thought superior to the Behrens-Fisher test (because of its greater power), and was defended by Bartlett, Barnard, Neyman and Pearson, among others.[16] Here there is just one relevant propensity, that of the pair of normal populations in question to yield sample pairs (x_1, x_2) with various ratios of sample standard deviations.

Fisher (1955) also gives the example of testing for association in a 2×2 contingency table with observed entries a, b, c, d. Since the marginal totals by themselves tells us nothing about association, Fisher's principle leads to a test conditional on the observed margins. Given the null hypothesis that the traits are independent, the probability of the observed table conditional on the observed margins is

$$\binom{a+b}{a}\binom{c+d}{c}\bigg/\binom{n}{a+c} = (a+b)!(a+c)!(b+d)!(c+d)!/\{n!a!b!c!d!\},$$

and Fisher's conditional test rejects the null hypothesis iff the probability that the first tabular entry, n_{11}, is as large or larger than the observed value, a, falls below the prescribed significance level. Here, too, there is an unconditional test with greater power but here, too, there is just one propensity, namely, that of the population to give random samples of n with different marginal totals. Still another example involving equality of Poisson means (and a single propensity) is discussed by Kalbfleisch and Sprott (1974), and they conclude (p. 108):

The various possible outcomes of an experiment sometimes vary considerably in the amount of information they provide concerning the hypothesis of interest. This is of no consequence in routine signalling situations, where only the long-run average informativeness is important. An outcome of below average informativeness will be compensated for by one that is above average at some time in the future. However, this will not be true in problems of inference, where one wishes to interpret the results of the experiment actually performed. If one unluckily obtains an uninformative outcome from which little or nothing can be learned ... no consolation will be derived from the fact that some future repetition, which may never by performed, could yield a more informative outcome to the experiment.

In problems of inference, the informativeness of the outcome actually observed must be taken into account. The data should be compared only with those outcomes which are of approximately the same informativeness or precision. In a test of significance, the observed outcome is compared with the other points of the reference set R. Therefore, R *should contain only those outcomes which are of approximately the same informativeness or precision as the one observed.*

Giere's admission that the intuitions of Fisher, Cox, Kalbfleisch, Sprott, and others are sound seems particularly damaging, given examples (like the Fisher-Behrens problem and association in a 2×2 table) which are resistant to the legerdemain (or propensity magic!) he applies to the examples of Cox and Basu and Birnbaum. A similar example is discussed in Jaynes (1976) which appertains to orthodox confidence intervals but teaches the same lesson (Jaynes, 1976, Example 5). Giere writes (p. 69):

I take it as undeniable that the intended point of these examples is sound. We should not use the mixed test.

But if those intuitions are undeniably sound when applied to Cox's example, are they not equally sound when applied to the examples of Fisher's I have cited, which are exactly like Cox's (who evidently modelled his examples after Fisher's) so far as the point under discussion is concerned? And the point, surely, is that it is unwise to base one's choice of significance test solely on considerations of size and power, as the Neyman-Pearson theory advises.

6. CONCLUSION

Many of the arguments of this chapter are familiar from the Bayesian literature, and I have tried to present them in a way that casts the underlying strategy into sharp relief. Thus, I have emphasized the extent to which internal criticism of orthodox methods has generated desiderata which are satisfied by Bayesian methods. The (mathematically determinate) lowering of the significance level with increasing sample size (Section 4) is a case in point. The demand that conditional tests (based on the outcome or process actually observed) be used in preference to mixed or unconditional tests (Section 5) is another.

I have also emphasized that Bayesian theory is simpler (i.e., more determinate) than orthodox theory. Heretofore, the orthodox statistician has reserved the mantle of objectivity for himself; yet, Bayesian methods leave less to chance and less to (unaided, unanalysed, personal) judgment.[17] Such judgments are required in selecting a significance level and sample size (or an appropriate trade-off between reliability and sensitivity), again in determining the rate at which the significance level should be decreased with increasing sample size, and even in the selection (Section 4) of the consequences of a theory to be tested. These judgments are ostensibly necessitated by the need to avoid (allegedly) subjective priors. Yet, judgmental probabilities are endemic to the judgments orthodox practice requires; these probabilities are, as Good (1976) emphasizes, 'swept under the carpet'. In particular, the judgment that there is no recognizable subset (Section 5) should be mentioned in this connection.

Finally, I have stressed the failure of orthodox theory to provide a satisfactory format for objective scientific reporting (or for conveying 'what the data have to tell us'). Attempts to flesh out the Neyman-Pearson concept of a statistical test (or 'rule of inductive behavior') were examined and found wanting. The next two chapters further articulate these criticisms.

NOTES

[1] Neyman (1950), p. 259.

[2] There is, to be sure, a (somewhat disreputable) tradition in philosophy that conceives belief as an 'act of will'. Pascal's Wager, for example, has us *deciding what to believe* on practical grounds. This is sheer confusion. I can decide how to behave (e.g., what rituals to follow), but I cannot literally decide what to believe (as opposed to opening myself to evidence, enquiring farther, and the like). Much ordinary talk lends itself to this confusion, as when we describe a jury as 'deliberating' and then 'reaching a verdict'. Ideally speaking, jurors are referees charged with judging whether the prosecution has marshalled sufficient evidence to establish guilt 'beyond a reasonable doubt'. Each juror registers his belief or disbelief that sufficient evidence has been mustered; the decision — the actual sentencing — is left to a judge, working within prescribed legal limits.

[3] Cf. Neyman *et al.* (1971) for the cloud-seeding experiments, and Neyman *et al.* (1956) for the work on the distribution of galaxies.

[4] They do not exist, e.g., for the important test of a simple hypothesis against a two-sided alternative. For any two-tailed test, the one-tailed test of the same size is more powerful against values of the parameter (e.g., a normal mean) to one side of the null value. For a thorough discussion of these matters, consult Lehmann (1959) or Kendall and Stuart (1967).

[5] E.g., Cramer (1946), Chapter 35.1.

[6] Cf. Anscombe (1963).

[7] Exceptions include G. A. Barnard (see his remarks in Savage *et al.* (1962), pp. 75ff.), Hacking (1965), Edwards (1972), Kalbfleisch and Sprott (1970).

[8] Savage *et al.* (1962), pp. 17–19 and Birnbaum (1962), (1969) contain useful introductions to this topic.

[9] E.g., we wish to test $p = 0.5$ vs. $p = 0.1$. Experiment 1 comprises 10 Bernoulli trials; experimenter 2 samples until the 2nd success is observed. The most powerful 5% test for the first experiment rejects $p = 0.5$ when fewer than 2 successes are observed, while that for the second experiment rejects when the 2nd success occurs after the 8th trials. If 2 successes are observed, therefore, and experimenter 2 observes the 2nd success on the 10th trial, $p = 0.5$ is accepted by experimenter 1 and rejected by experimenter 2.

[10] A very clear example occurs on p. 139 of Feller (1957). In testing the efficacy of a vaccine, the null hypothesis (of no affect) equates a vaccinated animal's chance of infection, p, with the normal rate of infection, which, in Feller's example, is 25%. Using the binomial law, one easily computes that 1 infection among 17 vaccinated is less probable than 0 in 10, both probabilities being conditional on the null hypothesis $p = 0.25$. Feller concludes: "It is therefore *stronger evidence* in favor of the serum if out of seventeen test animals only one gets infected than if out of ten all remain healthy". On the other hand, an *effective* vaccine for this case might be one which lowered the infection rate to 5% or less. Yet, for a test of $p = 0.25$ vs $p = 0.05$, 0 in 10 gives a higher likelihood ratio in favor of $p = 0.05$ than 1 in 17.

[11] See the articles by Sterling and Tullock in Morrison and Henkel (1970).

[12] The point is discussed in many of the articles reprinted in Morrison and Henkel (1970), for instance, that by David Bakan (Bakan, 1967).

[13] Lehmann (1959), Example 8, p. 23.

[14] Bakan (cf. Note 12), esp. p. 241.

[15] Lindley (1971) shows that fixing the significance level is also incoherent in the sense of generating an intransitive set of preferences among tests. The point is discussed in Giere (1976), pp. 97–98.

[16] The point is discussed in many places by R. A. Fisher; cf. e.g., Fisher (1955), (1956a), (1956b), and also Kalbfleisch and Sprott (1974), (1976).

[17] Example 3 of Jaynes (1976) provides a beautiful but somewhat technical illustration.

BIBLIOGRAPHY

Anscombe, F. J.: 1963, 'Tests of Goodness of Fit', *J. Roy. Stat. Soc.*, B, **25**, 81–94.

Bakan, D.: 1967, 'The Test of Significance in Psychological Research', *On Method*, Jossey-Bass, San Francisco, 1–29 (reprinted in Morrison and Henkel).

Barnard, G. A.: 1947, 'The Meaning of a Significance Level', *Biometrika* **34**, 179–182.

Birnbaum, A.: 1962, 'On the Foundations of Statistical Inference', *J. Am. Stat. Assoc.* **57**, 269–306.

Birnbaum, A.: 1969, 'Concepts of Statistical Evidence', *Philosophy, Science and Method* (S. Morgenbesser, P. Suppes, and M. White, eds.), St. Martin's Press, New York.

Cox, D. R.: 1958, 'Some Problems Connected With Statistical Inference', *Ann. Math. Stat.* **29**, 357–372.

Cramer, H.: 1946, *Mathematical Methods of Statistics*, Princeton University Press, Princeton, New Jersey.

Edwards, A. W. F.: 1972, *Likelihood*, Cambridge University Press, Cambridge.

Feller, W.: 1957, *Introduction to Probability Theory and its Applications*, Wiley, New York.

Fisher, R. A.: 1955, 'Statistical Methods and Scientific Induction', *J. Roy. Stat. Soc.*, B, **17**, 69–78 (reprinted in Fisher, 1974).

Fisher, R. A.: 1956a, 'On a Test of Significance in Pearson's *Biometrika* Tables (No. 10)', *J. Roy. Stat. Soc.*, B, **56–60** (reprinted in Fisher, 1974).

Fisher, R. A.: 1956b, *Statistical Methods and Scientific Inference*, Oliver and Boyd, Edinburgh.

Fisher, R. A.: 1974, *The Collected Papers of R. A. Fisher*, Vol. 5 (ed. by J. H. Bennett), University of Adelaide Press, Adelaide.

Giere, R. N.: 1976, 'Empirical Probability, Objective Statistical Methods, and Scientific Inquiry', *Foundations of Probability Theory, Statistical Inference, and Statistical Theories of Science* (W. A. Harper and C. K. Hooker, eds.), Vol. 2, pp. 63–101.

Good, I. J.: 1976, 'The Bayesian Influence, or How to Sweep Subjectivism Under the Carpet', *Foundations of Probability Theory, Statistical Inference and Statistical Theories of Science* (W. A. Harper and C. K. Hooker, eds.), Vol. 2, D. Reidel, Dordrecht–Holland.

Hacking, I.: 1965, *Logic of Statistical Inference*, Cambridge University Press, Cambridge.

Hodges, J. L. and Lehmann, E. L.: 1954, 'Testing the Approximate Validity of a Null Hypothesis', *J. Roy. Stat. Soc.*, B, **16**, 261–268.

Jaynes, E. T.: 1968, 'Prior Probabilities', *IEEE Trans. on Systems Science and Cybernetics*, SSC–4, 227–241.

Jaynes, E. T.: 1976, 'Confidence Intervals vs Bayesian Intervals', *Foundations of*

Probability Theory, Statistical Inference, and Statistical Theories of Science (W. A. Harper and C. K. Hooker, eds.), Vol. 2, D. Reidel, Dordrecht–Holland.

Kalbfleisch, J. D. and Sprott, D. A.: 1970, 'Applications of Likelihood Methods to Models Involving Large Numbers of Parameters', *J. Roy. Stat. Soc.*, B, **32**, 175–208.

Kalbfleisch, J. D. and Sprott, D. A.: 1974, 'On the Logic of Tests of Significance with Special Reference to Testing the Significance of Poisson-Distributed Observations', *Information, Inference, and Decision* (G. Menges, ed.), D. Reidel, Dordrecht–Holland.

Kalbfleisch, J. D. and Sprott, D. A.: 1976, 'On Tests of Significance', *Foundations of Probability Theory, Statistical Inference, and Statistical Theories of Science*, Vol. 2, D. Reidel, Dordrecht–Holland.

Kendall, M. G. and Stuart, A.: 1967, *The Advanced Theory of Statistics*, Vol. 2, Griffin, London.

Kyburg, H.: 1974, *The Logical Foundations of Statistical Inference*, Chapter II, D. Reidel, Dordrecht–Holland.

Kyburg, H. and Smokler, H. (eds.): 1964, *Studies in Subjective Probability*, Wiley, New York.

Lehmann, E. L.: 1958, 'Significance Level and Power', *Ann. Math. Stat.* **29**, 1167–1176.

Lehmann, E. L.: 1959, *Testing Statistical Hypotheses*, Wiley, New York.

Lieberman, B.: 1971, *Contemporary Problems in Statistics*, Oxford University Press, Oxford.

Lindley, D. V.: 1957, 'A Statistical Paradox', *Biometrika* **44**, 187–192.

Lindley, D. V.: 1971, *Bayesian Statistics, a Review*, SIAM, Philadelphia.

Marks, M: 1951, 'Two Kinds of Experiment Distinguished in Terms of Statistical Operations', *Psych. Rev.* **58**, 179–184 (reprinted in Lieberman).

Morrison, D. E. and Henkel, R. E.: 1970, *The Significance Test Controversy*, Aldine, Chicago.

Morton, N.: 1955, 'Sequential Tests for the Detection of Linkage', *Am. J. Hum. Genetics* 7, 277–318.

Neyman, J.: 1950, *First Course in Probability and Statistics*, Henry Holt, New York.

Neyman, J.: 1957, ' "Inductive Behavior" as a Basic Concept of Philosophy of Science', *Rev. Inst. Int. de Stat.* **25**, 7–22.

Neyman, J. and Pearson, E. S.: 1933, 'The Testing of Statistical Hypotheses in Relation to Probabilities *a priori*', *Proc. Cambridge Phil. Soc.* **29**, 492–510.

Neyman, J., Lovasich, J., Scott, E., and Wells, M.: 1971, 'Further Studies of the Whitetop Cloud-Seeding Experiment', *Proc. Nat. Acad. Sci,* **68**, 147–151.

Neyman, J., Scott, E., and Shaw, C. D.: 1956, 'Statistical Images of Galaxies with Particular Reference to Clustering', *Proc. Third Berkeley Symposium on Mathematical Statistics and Probability*, Vol. III, University of California Press, Berkeley.

Robbins, H.: 1963, 'A New Approach to a Classical Statistical Decision Problem', *Induction: Some Current Issues* (H. Kyburg and E. Nagel, eds.) Wesleyan University Press, Middletown, Conn.

Rozeboom, W.: 1960, 'The Fallacy of the Null Hypothesis Significance Test', *Psych. Bull.* **57**, 416–428 (reprinted in Morrison and Henkel).

Savage, L. J.: 1961, 'The Foundations of Statistics Reconsidered', *Proc. Fourth Berkeley Symposium on Mathematical Statistics and Probability*, University of California Press, Berkeley (reprinted in Kyburg and Smokler).

Savage, L. J., *et al.*: 1962, *The Foundations of Statistical Inference*, Methuen, London.

Spielman, S.: 1973, 'A Refutation of the Neyman-Pearson Theory of Testing', *Brit. J. Phil. Sci.* **24**, 201–222.

CHAPTER 11

BAYES/ORTHODOX COMPARISONS

1. INTRODUCTION

Our purpose in this chapter is to compare empirically the performance of Bayes and orthodox approaches to regression and related problems. Specifically, I treat the problem of identifying the degree of a polynomial and the order of a Markov chain. When we raise the degree of a polynomial or the order of a Markov chain, we improve the model's accuracy at the cost of some simplicity. According to the Bayesian analysis of Chapter 5, there is a well-defined rate of exchange that must be exceeded for the average likelihood (or support) of the model to increase. The exact average likelihoods are computable in both cases. It is worth stressing that the Bayesian approach to such problems is unified: one compares average likelihoods. By contrast, orthodox statistics offers us a mixed bag of tricks with no single (or simple) underlying logic.

2. THE NORMAL LINEAR MODEL

In the usual normal linear model for polynomial regression:

$$(11.1) \qquad Y_i = \beta_0 + \beta_1 X_i + \cdots + \beta_{k-1} X_i^{k-1} + u_i, \qquad i = 1, \ldots, n,$$

the error term u_i is assumed to be normally distributed about 0 with $E(u_i u_j) = 0$ for $i \neq j$ and $= \sigma^2$ for $i = j$. Thus, the error terms associated with different values of the independent variable X have the same variance (which I assume known) and zero covariance. If we write (11.1) in matrix notation:

$$(11.2) \qquad \mathbf{Y} = \mathbf{XB} + \mathbf{U}$$

with

$$\mathbf{Y} = \begin{bmatrix} Y_1 \\ \vdots \\ Y_n \end{bmatrix} \quad \mathbf{B} = \begin{bmatrix} \beta_0 \\ \vdots \\ \beta_{k-1} \end{bmatrix} \quad \mathbf{U} = \begin{bmatrix} u_1 \\ \vdots \\ u_n \end{bmatrix} \quad \mathbf{X} = \begin{bmatrix} 1 & X_1 & X_1 \ldots X_1^{k-1} \\ 1 & X_2 & X_2^2 \ldots X_2^{k-1} \\ \cdot & \cdot & \cdot \ldots \cdot \\ 1 & X_n & X_n^2 \ldots X_n^{k-1} \end{bmatrix},$$

where we assume rank $\mathbf{X} < n$, then writing $\hat{\mathbf{B}}$ for the column vector of least-squares estimates (which are also *ML* estimates), one easily obtains:

(11.3) $\hat{B} = (\mathbf{X}'\mathbf{X})^{-1}\mathbf{X}'\mathbf{Y}$, $\mathbf{X}'$ the transpose of $\mathbf{X}$.

Moreover,

(11.4) $\mathbf{X}'\mathbf{X} = \begin{bmatrix} n & \Sigma X_i & \Sigma X_i^2 \dots \Sigma X_i^{k-1} \\ \Sigma X_i & \Sigma X_i^2 & \Sigma X_i^3 \dots \Sigma X_i^k \\ \cdot & \cdot & \cdot \dots \cdot \\ \Sigma X_i^{k-1} & & \cdot \dots \Sigma X_i^{2k-2} \end{bmatrix}$,

and the likelihood can be written:

(11.5) $p(\mathbf{Y}/\mathbf{B}) \propto \exp[-2(2\sigma^2)^{-1}(\mathbf{B}-\hat{\mathbf{B}})'\mathbf{X}'\mathbf{X}(\mathbf{B}-\hat{\mathbf{B}})]$.

Hence, the likelihood is exactly normal. The informationless prior of $\mathbf{B}$ is locally uniform (Chapter 3), and so the posterior distribution of $\mathbf{B}$ is

(11.6) $p(\mathbf{B}/\mathbf{Y}) = (2\pi\sigma^2)^{-k/2}|\mathbf{X}'\mathbf{X}|^{1/2}\exp[-(2\sigma^2)^{-1}(\mathbf{B}-\hat{\mathbf{B}})'\mathbf{X}'\mathbf{X}(\mathbf{B}-\hat{\mathbf{B}})]$,

which is the multivariate normal density about $\hat{\mathbf{B}}$ with covariance matrix $\sigma^2(\mathbf{X}'\mathbf{X}^{-1})$. Because the likelihood is exactly normal, the approximation (Chapter 5)

(11.7) $A = \int\int \cdots \int p(\mathbf{Y}/\mathbf{B})\,d\mathbf{B} \doteq (2\pi)^{k/2}I^{-1/2}p(\mathbf{Y}/\hat{\mathbf{B}})$

gives an exact result. If, for convenience, we take $\sigma^2 = 1$, and write I_k for the determinant of the information matrix of the model (11.1), then, from (11.6):

(11.8) $I_k = |\mathbf{X}'\mathbf{X}|$,

with $\mathbf{X}'\mathbf{X}$ given by (11.4).

It is now a simple matter to compare the average likelihoods of any two polynomial models using (11.7) and (11.8). To simplify the calculations, I let the independent variable X run through the values $1, \dots, 6$. Then $\Sigma X = 21$, $\Sigma X^2 = 91$, $\Sigma X^3 = 441$, $\Sigma X^4 = 2275$, $\Sigma X^5 = 12201$, $\Sigma X^6 = 67171$, $\Sigma X^7 = 376761$, and $\Sigma X^8 = 2142595$. I indicate the inverse of $\mathbf{X}'\mathbf{X}$ below for the linear, quadratic, cubic and quartic cases of (11.1), corresponding to

$k = 2, 3, 4, 5$:

linear $(\mathbf{X'\,X})^{-1} = 105^{-1} \begin{bmatrix} 91 & -21 \\ -21 & 6 \end{bmatrix}$

quadratic $(\mathbf{X'\,X})^{-1} = 3920^{-1} \begin{bmatrix} 12544 & -7644 & 980 \\ -7644 & 5369 & -735 \\ 980 & -735 & 105 \end{bmatrix}$

cubic $(\mathbf{X'\,X})^{-1} = 254016^{-1} \begin{bmatrix} 3302208 & -3626784 & 1100736 & -98784 \\ -3626784 & 4287080 & -1352400 & 124264 \\ 1100736 & -1352400 & 438984 & -41160 \\ -98784 & 124264 & -41160 & 3920 \end{bmatrix}$

quartic

$(\mathbf{X'\,X})^{-1} = 20901888^{-1} \begin{bmatrix} 1588543488 & -2566287360 & 1310722560 \\ -2566287360 & 4258515456 & -2212648704 \\ 1310722560 & -2212648704 & 1166695488 \\ -264176640 & 451196928 & -240637824 \\ 18289152 & -31497984 & 16946496 \end{bmatrix}$

$\begin{bmatrix} -264176640 & 18289152 \\ 451196928 & -31497984 \\ -240637824 & 16946496 \\ 50109696 & -3556224 \\ -3556224 & 254016 \end{bmatrix}$

These matrices are all symmetric; the determinants are the quantities whose reciprocals appear outside, viz. 105, 3920, 254016, and 20901888. Finally, an easy matrix matrix multiplication gives:

$$(11.9) \qquad \mathbf{X'Y} = \begin{bmatrix} \Sigma Y \\ \Sigma XY \\ \vdots \\ \Sigma X^{k-1} Y \end{bmatrix} \qquad k = 2, 3, 4, 5.$$

This quantity must be recalculated for each new vector $\mathbf{Y}$ of observations in applying (11.3) to obtain the *ML* estimate $\hat{\mathbf{B}}$.

Writing I_k for the $k \times k$ matrix $\mathbf{X'X}$ associated with the polynomial (11.1) of degree $k - 1$, we see that the ratio of average likelihoods for polynomials of degrees $k - 1$ and $k - 2$ is $(2\pi)^{1/2} [I_{k-1}/I_k]^{1/2} \lambda(k-1; k-2)$, where $\lambda(k-1; k-2)$ is the ratio of the maximum likelihoods. Hence, the ratio exceeds one iff

$$(11.10) \qquad \lambda(k-1; k-2) > [I_k/I_{k-1}]^{1/2} (2\pi)^{-1/2} = \lambda^*(k-1; k-2).$$

The critical ratios $\lambda^*(k - 1; k - 2)$ for $k = 3, 4, 5$ are:

$$\lambda^*(2; 1) = (3920/105)^{1/2}(2\pi)^{-1/2} = 2.44 ,$$

$$\lambda^*(3; 2) = (254016/3920)^{1/2}(2\pi)^{-1/2} = 3.21 ,$$

$$\lambda^*(4; 3) = (20901888/254016)^{1/2}(2\pi)^{-1/2} = 3.62 .$$

Thus, for the quadratic to have higher average likelihood than the general linear polynomial requires that the ratio of their maximum likelihoods (i.e., the likelihoods of their best-fitting special cases) exceed 2.4, and so on. We go on increasing the degree of the polynomial for as long as the average likelihood goes on increasing. The maximum likelihoods themselves are given by:

$$(11.11) \qquad (2\pi)^{-n/2} \exp[-\Sigma e_i^2/2] ,$$

where the $e_i = Y_i - \hat{Y}_i$ are the *residuals*, viz., the deviations between the observations Y_i and the values $\hat{Y}_i$ predicted by the best-fitting polynomial of the given degree. Since the constant term $(2\pi)^{-n/2}$ cancels out of the ratios $\lambda(k - 1; k - 2)$, we need compute only the exponential factors of (11.11).

EXAMPLE 1. If the polynomial were $Y = (X - 1)^2$ error-free observations at $X = 1, 2, 3, 4, 5, 6$, would give $Y = 0, 1, 4, 9, 16, 25$. Entering a table of random normal deviates, I obtained the 6-vector $(-1.45, 0.33, 0.87, -1.90, 0.69, -0.36)$. The corresponding random sample from $Y = (X - 1)^2$ is $(-1.45, 1.33, 4.87, 7.10, 16.69, 24.64)$. For (11.9) I calculate $\Sigma Y = 53.18$, $\Sigma XY = 275.51$, $\Sigma X^2 Y = 1465.59$ and from these entries of $\mathbf{X'Y}$ I quickly obtain the *ML* polynomials of degrees $1-4$:

$$linear \ \hat{\mathbf{B}} = 105^{-1} \begin{bmatrix} 91 & -21 \\ -21 & 6 \end{bmatrix} \begin{bmatrix} 53.18 \\ 275.51 \end{bmatrix} = \begin{bmatrix} -9.0127 \\ 5.1074 \end{bmatrix}$$

and so the best-fitting line is $-9.0127 + 5.1074X$, with $\Sigma e^2 = 35.83$, and likelihood kernel $\exp[-35.83/2] = 1.65 \times 10^{-8}$.

$$quadratic \ \hat{\mathbf{B}} = \begin{bmatrix} -0.671 \\ 1.149 \\ 0.894 \end{bmatrix}$$

found similarly, and so the best-fitting quadratic is $-0.671 +$

$1.149X + 0.894X^2$, with $\Sigma e^2 = 6.01$ and likelihood kernel $\exp[-6.01/2] = 0.049$.

cubic: The best-fitting cubic is $-3.940 + 2.963X - 0.468X^2 + 0.130X^3$, with $\Sigma e^2 = 4.90$ and likelihood kernel $\exp[-4.9/2] = 0.086$;

quartic: the best-fitting (*ML*) quartic is $-14.335 + 20.866X - 10.100X^2 + 2.151X^3 - 0.144X^4$, with $\Sigma e^2 = 3.58$ and likelihood kernel $\exp[-3.58/2] = 0.167$. Upon taking *ML* ratios I get: $\lambda(2; 1) = 2966392.5$, $\lambda(3; 2) = 1.7$, and $\lambda(4; 3) = 1.9$. Only the first of these exceeds the corresponding critical value λ^*, and so the Bayes method correctly identifies the polynomial as quadratic.

3. ORTHODOX REGRESSION ANALYSIS

Analysis of variance techniques, with which I assume some familiarity, all turn on the possibility of partitioning a sum of squares. In regression, the total variation of the observations partitions into a sum of two terms, one the variation of the Y_i about their mean $\overline{Y}$, called the *explained sum of squares* (or *explained* **SS**), and the other, the *residual* **SS**, Σe^2:

$$(11.12) \quad \Sigma(Y_i - \overline{Y}_i)^2 = \Sigma(\hat{Y}_i - \overline{Y})^2 + \Sigma(\hat{Y}_i - Y_i)^2,$$

where $\Sigma(\hat{Y}_i - Y_i)^2 = \Sigma e^2$ It will prove convenient to put these quantities in matrix form for computational purposes. First, $\Sigma(Y_i - \overline{Y})^2 = \Sigma Y_i^2 - n^{-1}(\Sigma Y_i)^2$, or, in matrix notation:

(A) $\qquad \Sigma(Y_i - \overline{Y})^2 = \mathbf{Y'Y} - n^{-1}(\Sigma Y_i)^2$ [Total *SS*].

Likewise, $\Sigma e^2 = \mathbf{E'E} = (\mathbf{Y} - \mathbf{X\hat{B}})'(\mathbf{Y} - \mathbf{X\hat{B}}) = \mathbf{Y'Y} - 2\mathbf{\hat{B}'X'Y} + \mathbf{\hat{B}'X'X\hat{B}}$, whence

(B) $\qquad \mathbf{E'E} = \mathbf{Y'Y} - \mathbf{\hat{B}'X'Y}$ [Residual *SS*]

since $\mathbf{X'X\hat{B}} = \mathbf{X'Y}$ by (11.3). Subtracting (B) from (A) gives

(C) $\qquad \Sigma(Y_i - \overline{Y})^2 = \mathbf{\hat{B}'X'Y} - n^{-1}(\Sigma Y_i)^2$ [Explained *SS*]

The *correlation coefficient* is defined as the ratio of the explained *SS* to the total *SS*:

$$(11.13) \quad R^2 = \frac{\mathbf{\hat{B}'X'Y} - (\Sigma Y)^2/n}{\mathbf{Y'Y} - (\Sigma Y)^2/n}.$$

It can be shown (Johnston, 1972, p. 128) that $E[\Sigma e^2] = (n - k)\sigma^2$, where, as before, σ^2 is the common variance of the error terms u_i, and $k - 1$ is the degree of the fitted polynomial, so that $\Sigma e^2/(n - k)$ is an unbiased estimator of σ^2 and $\Sigma e^2/\sigma^2$ has a chi square distribution with $n - k$ degrees of freedom (df). Also, $E[\Sigma(Y_i - \overline{Y})^2] = (n - 1)\sigma^2$, so that $\Sigma(Y_i - \overline{Y})^2/(n - 1)$ is an unbiased estimator of σ^2 and $\Sigma(Y_i - \overline{Y})^2/\sigma^2$ has a chi square distribution with $n - 1$ df, which, moreover, is independent of the distribution of $\Sigma e^2/\sigma^2$. Now we can express the correlation coefficient in the equivalent form $R^2 = 1 - (\Sigma e^2/n)/(\Sigma y^2/n)$, where $y_i = Y_i - \overline{Y}$. If we replace $\Sigma e^2/n$ and $\Sigma y^2/n$ by the corresponding unbiased estimators, $\Sigma e^2/(n - k)$ and $\Sigma y^2/(n - 1)$, we obtain a correlation coefficient adjusted for degrees of freedom:

$$(11.14) \qquad R_{adj}^2 = 1 - \frac{\Sigma e^2/(n - k)}{\Sigma y^2/(n - 1)}.$$

As we raise the degree of the fitted polynomial, R^2, like the maximum likelihood, will always increase, but, like the average likelihood, the adjusted correlation coefficient (adjusted, that is, for degrees of freedom) needn't increase. The *method of adjusted correlation coefficients* identifies the degree of the fitted polynomial as that for which R_{adj}^2 assumes its maximum.

That R^2 increases means that the proportion of the total variation explained by the fitted polynomial increases with its degree. In particular, the explained SS increases with the degree of the polynomial, but, beyond some point, at a diminishing rate. This suggests adding a new regressor X^k to $X, X^2, \ldots, X^{k-1}$ iff the mean SS due to X^k is large compared to the mean residual SS after adding X^k. A *mean* SS is the SS divided by its df and gives, as we have seen, an unbiased estimate of σ^2. Since mean SS's have chi squared distributions, ratios of them have an F-distribution. The *mean SS due to X^k* is the explained SS of $X, X^2, \ldots, X^k$ less that of $X, X^2, \ldots, X^{k-1}$ and has one df. Hence X^k is added iff $F = $ (mean SS due to X^k)/(residual SS after adding X^k) is improbably large as determined by the tables of the F-distribution with $(1, n - k)$ df. This method is called *stepwise regression*.

EXAMPLE 1 (concluded). First, I compute the adjusted correlation coefficients. From (A) above, $y^2 = 492.3319$ (total SS). The residual SS's were found above to be 35.83, 6.01, 4.90, and 3.58. Inserting the appropriate values and df in (11.14), $k - 1$ being the degree of the fitted polynomial, we obtain $R_{adj}^2 = 0.909, 0.980, 0.975$, and 0.964 for $k - 1 = 1, 2, 3, 4$. Hence the method of adjusted correlation coefficients also correctly identifies the polynomial as quadratic.

TABLE I

Source of variation	Explained SS	df	Mean SS
X	456.3249	1	
addition of X^2	29.9312	1	29.9312
X and X^2	486.2561	2	
residual SS	6.0758	3	2.0253
total SS	492.3319	5	

$F = 14.78$

TABLE II

Source of variation	Explained SS	df	Mean SS
X and X^2	486.2561	2	
addition of X^3	0.9462	1	0.9462
X, X^2 and X^3	487.2923	3	
residual SS	5.0396	2	2.5198
total SS	492.3319	5	

$F = 0.37$

Stepwise regression is best set out in tabular form. I evaluate first the effect of adding X^2 to X. The residual SS after adding X^2 is got by subtracting the explained SS for X, X^2 from the total SS of 492.3319. We obtain $F = 29.9312/2.0253 = 14.78$, which is, at $(1, 3)$ df, significant at the 5% level, though not at the 1% level. I evaluate next the addition of X^3. Here, too, the explained SS for X, X^2 and X^3 is obtained from (C) above, and the SS due to X^3 is gotten by subtraction. Now $F = 0.9462/2.5198 = 0.37$ at $(1, 2)$ df, which is clearly not significant at the 5% level. Hence, the analysis proceeds no farther, and so, at the 5% level (though not at the more conservative 1% level), the degree of the polynomial is correctly identified. It cannot be overemphasized, there is no way of telling which choice of significance level will produce the smallest overall probability of misidentifying the degree of the polynomial. I ran the orthodox procedure at the 1%, 5%, and 10% levels.

4. THE RESULTS

I sampled several polynomials of degrees 1–4 and compared the percentages of correct identifications (of the degree) scored by the Bayes method and the

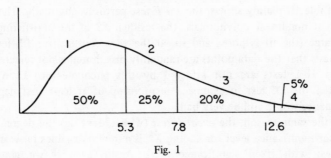

Fig. 1

F-test at 10%, 5% and 1% significance levels. I observed the independent variable at $X = 1, 2, 3, 4, 5, 6$ by generating 6-tuples of random normal deviates and computing the observed Y values (cf. Example 1). The sum of the squares of 6 random normal deviates has, of course, the chi square distribution with 6 df. I used the (upper) 50%, 75%, and 95% points of this distribution to apportion the 6-tuples among four regions of increasing deviance (cf. Figure 1, where the % of the total probability mass concentrated in each region is also shown). Twenty 6-tuples were generated in each region, the % correct for each method scored, and then the weighted average $0.5p_1 + 0.25p_2 + 0.2p_3 + 0.05p_4$ used as an overall estimate of accuracy, p_i being the % correct in region i, $i = 1, \ldots, 4$. E.g., for the quadratic $(X - 1)^2$ of Example 1, the Bayes method identified the following numbers of 6-tuples (out of a possible 20) correctly (as quadratic) in the four regions 1–4, resp.: 20, 14, 16, 15. Its overall estimated accuracy for this polynomial is therefore $0.5(100\%) + 0.25(70\%) + 0.2(80\%) + 0.05(75\%) = 87.25\%$.

One might wonder whether twenty 6-tuples per region is enough to provide a reasonably reliable estimate of accuracy. To check this, I generated another set of eighty 6-tuples (20 per region). The results are shown in Table III, and strongly indicate that the estimate based on 20 per region is quite reliable. I used my original set of eighty throughout, therefore, varying the polynomial.

TABLE III

	Bayes	10% F-test	5% F-test	1% F-test
Sample 1	87.25	86.5	94.25	56.25
Sample 2	87.75	88.5	91.75	59.25
Combined	87.5	87.5	93.0	57.75

As Table III plainly shows, the 1% F-test performs abysmally. The reason is that, at non-linear polynomials, the residual SS of the best-fitting line is often large (the fit is poor) and so, at the very conservative 1% level, the hypothesis that the data points are randomly distributed is not rejected. (The 5% and 10% tests are not affected by this phenomenon.) Even if one permitted the 1% test to assume a relationship of at least first degree, its performance would still be very bad.

For the same reason, the *exact F-test* (which identifies the degree as k iff the exact significance level for adding X^k is a minimum) does far worse than an F-test with fixed significance level. Again, even if we ignore the significance level for adding X, the exact F-test fares badly. In sampling from a cubic, for example, the $P(F > F^*)$ for adding X^2 are slightly but consistently smaller than those for adding X^3. The estimated accuracies of the four methods for a dozen polynomials are shown in Table IV.

TABLE IV

	Bayes	10% F-test	5% F-test	1% F-test
X	88.5	91.5	86.75	49.0
$(X - 1)^2$	87.25	86.5	94.25	56.25
$0.01X^2 + X$	7.5	5.25	3.75	0
X^3	87.0	83.75	89.25	34.75
$(X - 1)^3$	87.75	91.5	90.75	0
$0.01X^3 + X^2$	11.25	10.75	4.75	2.5
$0.001X^3 - X^2$	12.5	10.75	4.75	3.75
$2X^3 - 6X^2 + 6X - 3$	87.75	91.5	96.25	0
X^4	87.5	80.75	52.0	0
$0.5X^4 - 4X^2 + 8$	·87.75	51.5	14.5	0
$0.01X^4 + X^3 - 2X^2$	16.0	11.25	6.0	0
$0.1X^4 - 3X^3 + X^2 - X$	24.25	7.0	5.0	2.5

These results powerfully suggest the overall superiority of the Bayes method: it never does much worse than the best orthodox test, and often fares much better. One is further struck by the consistency of its performance. Unlike the orthodox methods, its accuracy does not appreciably diminish at higher degree polynomials. Thus, the Bayes method is close to 90% accurate at each of X, $(X - 1)^2$, X^3 and X^4. But the orthodox tests drop off steeply between the cubic and the quartic, and this despite the fact that they could not commit errors of overestimation for the quartic (while the Bayes method was allowed to commit such errors). By raising the significance level with the degree of the polynomial, the *F*-test improves somewhat. But, of course, one does not know the degree of the polynomial in practice. Finally, the performance of the Bayes test also falls off less sharply as the leading coefficient is reduced, rendering the given polynomial less discriminable from polynomials of lower degree. The data suggest, in short, that, for *some* choice of significance level, the *F*-test can do about as well as the Bayes test. But you can never know (and seldom guess) which significance level will deliver optimal or even satisfactory performance without knowing the answer to the question posed. And, it goes without saying, no meaningful power functions can be computed for this problem.

5. HIGHER ORDER MARKOV CHAINS

In a Markov chain M_r of order r, the probabilities of current trial outcomes depend on the outcomes of the r previous trials. My assumption throughout this section is that transition probabilities are stationary. The treatment will be confined to dichotomous processes: at each trial, the process can be in one of two states, labelled '*a*' and '*b*'. Note that by my convention, M_0 is the general *random stationary* (or Bernoulli) process, while M_1 is the general 'Markov chain' in the usual sense.

My objective is to compare the accuracy of the Bayes test for the order of a Markov chain with that of the orthodox tests given in Section 3.3 of Anderson and Goodman (1957). The procedure is simple and exact. The computer runs through all 2^n dichotomous sequences of length n. We obtain the accuracy of a given method by toting the probabilities of correctly identified sequences, where these probabilities are conditional on the particular Markov process of order r to be identified. (Just as in the regression case, we expect an accurate test to correctly identify 'common' or probable sequences and misidentify only the more 'deviant' sequences.) In this section, I describe and illustrate the tests to be compared.

Let n_i be the number of occurrences of state i, $|n_{ij}|$ the number of occurrences of the digram ij, etc., and let p_i, $p_{ij}, \ldots$ be the corresponding transition probabilities for $M_0, M_1 \ldots$. Their *ML* estimates are:

$$(11.15) \qquad p_i = n_i/n, \qquad p_{ij} = n_{ij}/n_i, \qquad p_{ijk} = n_{ijk}/n_{ij}, \ldots$$

N.B., the term n_i in the denominator of p_{ij} counts the number of steps or trials in state i *prior to the last trial*, and likewise for n_{ij}, n_{ijk}, etc., when these appear as denominators. For in estimating p_{ij}, we want the proportion of times the process moves from i to j but if i occurs on the last trial of the sequence, the process *cannot* move to state j.

The relevant average likelihoods, based on a uniform prior, are easily (and exactly) obtained. Consider, for illustration, M_1. The probability of a given sequence is:

$$(11.16) \qquad p_{aa}^{n_{aa}} p_{ab}^{n_{ab}} p_{ba}^{n_{ba}} p_{bb}^{n_{bb}} \times 2^{-1},$$

assuming that both states have equal probability of occurring on the first trial (an assumption I retain throughout). The average likelihood is just the integral of (11.16), and noting that the (unknown) transition probabilities satisfy $p_{ab} = 1 - p_{aa}$, $p_{ba} = 1 - p_{bb}$, while $n_{aa} + n_{ab} = n_a$, $n_{bb} + n_{ba} = n_b$, the integral is a Beta function equal to:

$$(11.17) \qquad [n_{aa}! n_{ab}! / (n_a + 1)!] \, [n_{ba}! n_{bb}! / (n_b + 1)!].$$

The average likelihoods for M_0, M_2, M_3, M_4 are similarly obtained and listed below:

$$\begin{aligned}
&M_0: n_a! n_b! / (n + 1)!, \\
&M_1: \text{Given by } (11.17) \times 2^{-1}, \\
&M_2: [n_{aaa}! n_{aab}! / (n_{aa} + 1)!] \, [n_{aba}! n_{abb}! / (n_{ab} + 1)!] \\
&\qquad \times [n_{baa}! n_{bab}! / (n_{ba} + 1)!] \, [n_{bba}! n_{bbb}! / (n_{bb} + 1)!] \times 2^{-2}, \\
&M_3: [n_{aaaa}! n_{aaab}! / (n_{aaa} + 1)!] \\
&\qquad \ldots [n_{bbba}! n_{bbbb}! / (n_{bbb} + 1)!] \, \times 2^{-3}, \\
&M_4: [n_{aaaaa}! n_{aaaab}! / (n_{aaaa} + 1)! \\
&\qquad \ldots [n_{bbbba}! n_{bbbbb}! / (n_{bbbb} + 1)!] \times 2^{-4}.
\end{aligned}$$

(11.18)

These formulas are easy to remember; the second half of each one is obtained from the first half by interchange of a and b. Also, there are 2^r factors for M_r, and the numerators bear an obvious relation to the associated denominator. The factor 2^{-r} in the expression for M_r is contributed by the initial r-gram with which the sequence starts (given our assumption that all possible initial r-grams are equiprobable). The formulas for the average likelihoods in the multi-state case are direct extensions of (11.18).

Anderson and Goodman (*op. cit.*) give two closely related chi square tests. In applying the Bayes test, one identifies the process as M_r iff the average likelihood for M_r is a maximum. In applying the two orthodox tests, we test, successively, the null hypotheses M_0, M_1, M_2, etc., that is, we test M_0 vs M_1, and then, if M_0 is rejected, go on to test M_1 vs M_2, and so forth, identifying the process as M_r iff M_r is the first unrejected hypothesis in this sequence. In testing, e.g., M_1 vs M_2, we are testing whether the current trial outcome depends on the preceding trial only or on the two preceding trials. Therefore, we are testing the four equalities:

$$p_{aaa} = p_{baa}, \quad p_{aab} = p_{bab}, \quad p_{bbb} = p_{abb}, \quad p_{bba} = p_{aba}.$$

Given M_1, the expected counts are therefore $n^*_{ijk} = n_{ij}\hat{p}_{jk}$ while on M_2 the expected counts are $n^*_{ijk} = n_{ij}\hat{p}_{ijk}$. The statistic

$$(11.19) \quad \chi^2_a = n_{aa}(\hat{p}_{aaa} - \hat{p}_{aa})^2/\hat{p}_{aa} + n_{aa}(\hat{p}_{aab} - \hat{p}_{ab})^2/\hat{p}_{ab}$$
$$+ n_{ba}(\hat{p}_{baa} - \hat{p}_{aa})^2/\hat{p}_{aa} + n_{ba}(\hat{p}_{bab} - \hat{p}_{ab})^2/\hat{p}_{ab}$$

measures the dependence of transitions from state **a** on trial $n + 1$ on trial n, and is associated with the 2×2 contingency table with entries $n_{aaa}, n_{aab}, n_{baa}, n_{bab}$. It follows that χ^2_a has (asymptotically) a chi square distribution with 1 degree of freedom. By interchange of 'a' and 'b' in χ^2_a we obtain a statistic χ^2_b, which measures the dependence of transitions from state **b** on trial $n + 1$ on trial n. Hence, the appropriate test statistic is

$$(11.20) \quad \chi^2 = \chi^2_a + \chi^2_b$$

which has (approximately) the chi square distribution with 2 df. In testing M_{r-1} vs M_r, the number of associated contingency tables is 2^{r-1}, and that is the number of degrees of freedom of the associated chi square statistic. This statistic then partitions into 2^{r-1} parts, as in the special case $r = 2$ of (11.20). Thus, for testing M_2 vs M_3, we use $\chi^2 = \chi^2_{aa} + \chi^2_{ab} + \chi^2_{ba} + \chi^2_{bb}$, and so forth.

The other orthodox test (illustrated here for M_1 vs M_2) is based on the modified ratio of maximum likelihoods:

$$(11.21) \quad \lambda_a = \prod_{i, k} (\hat{p}_{ak}/\hat{p}_{iak})^{n_{iak}}$$

and this gives rise to the statistic $\chi^2_a = -2\ln \lambda_a$, which has, asymptotically, a chi square distribution with 1 df. Define λ_b from λ_a by interchange of 'a' and 'b' and set $\chi^2_b = -2\ln \lambda_b$. Then $\chi^2 = \chi^2_a + \chi^2_b$ has (approximately) a chi square distribution with 2 df and can be used to test M_1 vs M_2. The extension to the general case, M_{r-1} vs M_r, is clear.

In applying the orthodox tests, the significance level is fixed in advance (no doubt, more elaborate procedures could be devised, but it is not clear that they would improve the performance of the test). One additional remark is necessary. Any of the denominators in (11.19) could vanish, and this raises a difficulty whose solution is by no means obvious. I will follow here the advice of Bradley Efron (oral communication) and set the entire term (in which a zero denominator occurs) equal to zero and without changing the *df*. The same thing can happen in (11.21), and in that case, I will set the entire term equal to one without altering the *df*. It is in general unclear how the possibility of small frequency counts affects the accuracy of the chi square approximation to the actual distribution of our statistics. None of these difficulties arise for the Bayes test (which is exact).

EXAMPLE 2. The process is second order with transition matrix:

	aa	ab	ba	bb
aa	0.2	0.8	0	0
ab	0	0	0.1	0.9
ba	0.5	0.5	0	0
bb	0	0	0.7	0.3

Following customary practice, I represent the passage from *aa* at the two preceding trials to *a* on the current trial as the passage from *aa* to *aa*, etc. (The passage from *ij* to *km* has zero probability for $j \neq k$.) In this way, a second order chain with two states a, b is represented, formally at least, as a first order chain with four states *aa*, *ab*, *ba*, *bb*. In the earlier notation we have $p_{aaa} = 0.2, p_{aab} = 1 - p_{aaa} = 0.8$, and so on. Consider the following sequence of length $n = 10$:

baabbaabbb .

The associated matrix of *ML* estimates $\hat{p}_{ijk}$ is:

	aa	ab	ba	bb
aa	0	1	0	0
ab	0	0	0.2	0.8
ba	0.8	0.2	0	0
bb	0	0	0.75	0.25

The probability of the given sequence (conditional on the given second order process) is $(0.5^2)(0.5)(0.8)(0.9)(0.7)(0.5)(0.8)(0.9)(0.3) = 0.006804$. Note that the probability has eight factors (excluding the probability, 0.5^2, of the initial digram). I list the average likelihoods of the M_r for this sequence below:

M_0: $4!6!/11! = 0.000433$,

M_1: $(2!2!/5!)(2!3!/6!)(0.5) = 0.000278$,

M_2: $(0!2!/3!)(0!2!/3!)(2!0!/3!)(1!1!/3!)(0.5^2) = 0.001543$,

M_3: $(0!0!/1!)(0!2!/3!)(0!0!/1!)(1!1!/3!)(0!2!/3!)(0!0!/1!)(1!0!/2!)$
$\quad (0!0!/1!)0.5^3 = 0.001157$,

M_4: $(0!0!/1!)(0!0!/1!)(0!0!/1!)(1!1!/3!)(0!0!/1!)(0!0!/1!)(1!0!/2!)$
$\quad (0!0!/1!)(0!0!/1!)(0!2!/3!)(0!0!/1!)(0!0!/1!)(1!0!/2!)(0!0!/1!)$
$\quad (0!0!/1!)(0!0!/1!)0.5^4 = 0.000868$.

Since the maximum occurs at M_2, the sequence in question is correctly identified as (having come from) an M_2 process. Hence, we enter the probability of this sequence, 0.006804, in the sum which gives the accuracy of the Bayes method in identifying the order of the given (second order) Markov process.

For the first orthodox test, I start by testing M_0 vs M_1:

$$\chi^2 = n_a(\hat{p}_{aa} - \hat{p}_a)^2/\hat{p}_a + n_a(\hat{p}_{ab} - \hat{p}_b)^2/\hat{p}_b + n_b(\hat{p}_{bb} - \hat{p}_b)^2/\hat{p}_b$$
$$+ n_b(\hat{p}_{ba} - \hat{p}_a)^2/\hat{p}_a$$
$$= 4(0.5 - 0.4)^2/0.4 + 4(0.5 - 0.6)^2/0.6 + 6(0.6 - 0.6)^2/0.6$$
$$+ 6(0.4 - 0.4)^2/0.4$$
$$= 0.166667 .$$

Here $r = 1$, and the df is therefore $2^{r-1} = 1$. Since the upper 5% point of the chi square distribution with 1 df is 3.81, M_0 is not rejected at the 5% level. Hence, the first orthodox test misidentifies the sequence as an M_0 (as random stationary). The second orthodox test also does so.

EXAMPLE 3. The homogeneous sequence *aaaaaaaaaa* poses an interesting test case for the rule-of-thumb (Chapter 5) that the simpler of two equally agreeing hypotheses is better supported. The homogeneous sequence can be regarded as random stationary with state **a** having probability 1, or as a Markov chain (M_1) with $P(a) = P(b)$ but $p_{aa} = 1$, or as a second order Markov chain with $p_{aaa} = 1$, etc. Clearly, each member of the sequence $M_0, M_1, M_2,$

... is simpler (in my sense) than its immediate successor. Hence, we should expect the Bayes test to classify homogeneous sequences as random stationary. The exact average likelihoods do decrease in the expected way: 1/11 (M_0), 1/20 (M_1), 1/36 (M_2), etc. The same result is obtained using the orthodox tests. In fact, $\chi^2 = 0$ for the test of M_0 vs M_1, using the orthodox test.

6. THE RESULTS

I said earlier that in applying the orthodox tests, we test M_0 against M_1, then M_1 against M_2, and so forth. This procedure parallels stepwise regression. But there is a way to improve the performance of the orthodox tests. For we can also test M_0 against M_r, for values of r greater than 1. E.g., to test M_0 against M_2, by the second orthodox method, say, one uses modified forms of the statistics λ_a and λ_b with the numerators appropriate for M_0 vs M_1 and the denominators those for testing M_1 vs M_2:

$$(11.22) \qquad \lambda_a = (\hat{p}_a/\hat{p}_{aaa})^{n_{aaa}}(\hat{p}_a/\hat{p}_{aba})^{n_{aba}}(\hat{p}_a/\hat{p}_{baa})^{n_{baa}}(\hat{p}_a/\hat{p}_{bba})^{n_{bba}},$$

with λ_b obtained from λ_a by the usual interchange of 'a' and 'b'. The statistics for testing M_0 versus the other M_r are defined similarly, and $\chi^2 = -2ln\lambda_a - 2ln\lambda_b$ has (approximately) a chi square distribution with $2^r - 1$ degrees of freedom (for the test of M_0 vs M_r). Now the modified procedure begins by testing M_0 against M_r for $r = 1, 2, 3, 4$, stopping at the least r for which M_0 is rejected. M_r is then tested against M_{r+1}, as in the original stepwise procedure, and the process terminates with the last unrejected M_r. If M_0 is not rejected against any M_r, the sequence is classified as random stationary.

I have found that the original stepwise procedure is far too conservative in practice. For instance, at the 5% significance level, almost all sequences of length 10 are classified as M_0 and none are classified higher than M_1. Of course, with sequences this short, one should use a less stringent significance level, e.g., 10% or even 20%. As in the regression case, though, the optimal significance level for sequences of given length is anybody's guess. By following the more elaborate procedure just outlined, one can improve the sensitivity of the test without raising the significance level. For many sequences show marked second order dependencies without exhibiting strong first order dependencies. In such cases, M_0 will be rejected in a test against

M_2, though not in a test against M_1. I found that the two orthodox tests differ inappreciably in their accuracy, and so I confined my final test to the second orthodox method, using significance levels of 5%, 10% and 20% and sequences of length 14. I appraised the accuracy of each of the methods for identifying a single Markov process of each of the orders $0, 1, 2, 3, 4$. This entails running through all $2^{14} = 16384$ binary sequences of length 14, and grouping (for each method of identification) the sequences classed as M_0, those classed as M_1, and so forth. (Fortunately, this only needs to be done once.) One then sums the probabilities of the sequences in each group (conditional on the Markov process to be identified). If the 'true' process is M_r, the sum of the probabilities of the sequences classed as M_r gives the accuracy of the method; the other sums give its probabilities of misidentifying the M_r process as an M_u process, for $u \neq r$. I list below the transition probabilities for the five Markov processes used.

Test 0.   The process is M_0 with $p_a = 0.3$.

Test 1.   The process is M_1 with $p_{aa} = 0.3, p_{bb} = 0.2$.   N.B., $p_{ab} = 1 - p_{aa}$, etc.

Test 2.   The process is M_2 with transition probabilities as in Example 2.

Test 3.   The process is M_3 with transition probabilities: $p_{aaaa} = 0.3$, $p_{aaba} = 0.5, p_{abaa} = 0.7, p_{abba} = 0.2, p_{baaa} = 0.6, p_{baba} = 0.4$, $p_{bbaa} = 0.2, p_{bbba} = 0.9$ (where, again, $p_{aaab} = 1 - p_{aaaa}$, etc.).

Test 4.   The process is M_4 with transition probabilities: $p_{aaaaa} = 0.1$, $p_{aaaba} = 0.6, p_{aabaa} = 0.4, p_{aabba} = 0.3, p_{abaaa} = 0.5$, $p_{baaaa} = 0.9, p_{baaba} = 0.2, p_{babaa} = 0.7, p_{babba} = 0.8$, $p_{bbaaa} = 0.15, p_{ababa} = 0.95, p_{abbaa} = 0.25, p_{abbba} = 0.45$, $p_{bbaba} = 0.35, p_{bbbaa} = 0.65, p_{bbbba} = 0.75$.

The results are given in Tables V–IX below, where the summed probabilities of sequences classified M_r are shown for each method.

TABLE V

Test 0.	M_0	M_1	M_2	M_3	M_4
Bayes	0.36	0.16	0.11	0.14	0.23
Orth 5%	0.92	0.05	0.03	0.00	0.00
Orth 10%	0.80	0.11	0.07	0.02	0.00
Orth 20%	0.58	0.23	0.14	0.05	0.00

The Bayes test identifies a random stationary sequence as such with probability 0.36, and identifies such a sequence erroneously as a first order chain with probability 0.16, etc., and similarly for the other methods.

TABLE VI

Test 1	M_0	M_1	M_2	M_3	M_4
Bayes	0.27	0.32	0.12	0.11	0.18
Orth 5%	0.85	0.12	0.03	0.00	0.00
Orth 10%	0.74	0.18	0.07	0.01	0.00
Orth 20%	0.51	0.29	0.16	0.04	0.00

TABLE VII

Test 2	M_0	M_1	M_2	M_3	M_4
Bayes	0.04	0.02	0.48	0.18	0.28
Orth 5%	0.59	0.01	0.40	0.00	0.00
Orth 10%	0.41	0.02	0.56	0.01	0.00
Orth 20%	0.21	0.03	0.72	0.04	0.00

TABLE VIII

Test 3	M_0	M_1	M_2	M_3	M_4
Bayes	0.05	0.05	0.05	0.42	0.43
Orth 5%	0.93	0.03	0.01	0.03	0.00
Orth 10%	0.69	0.09	0.06	0.16	0.00
Orth 20%	0.40	0.19	0.11	0.30	0.00

TABLE IX

Test 4	M_0	M_1	M_2	M_3	M_4
Bayes	0.07	0.08	0.16	0.16	0.53
Orth 5%	0.82	0.07	0.10	0.01	0.00
Orth 10%	0.65	0.12	0.21	0.02	0.00
Orth 20%	0.45	0.19	0.30	0.06	0.00

The M_n column of the table for test n gives the accuracy of the different methods for test n (i.e., for identifying the given Markov process of order n, $n = 0, 1, 2, 3, 4$). Table X gives these results.

TABLE X

	Test 0	Test 1	Test 2	Test 3	Test 4
Bayes	0.36	0.32	0.48	0.42	0.53
Orth 5%	0.92	0.12	0.40	0.03	0.00
Orth 10%	0.80	0.18	0.56	0.16	0.00
Orth 20%	0.58	0.29	0.72	0.30	0.00

The results exhibit a clear pattern that parallels the regression case. The orthodox tests break down completely at M_4 even when we set the significance level as high as 20%. Even at this high a level, the orthodox test has higher probability of identifying a sequence drawn at random from an M_r process as random stationary than as $M_r, r = 1, 2, 3, 4$. By contrast, the Bayes test has higher probability of identifying an M_r sequence as such than as any $M_u, u \neq r$. (The single apparent exception at M_3 — Table VIII — is due to the fact that we arbitrarily disallowed sequences to be classed as higher than fourth order.) The conservatism of the stepwise orthodox procedure described above is even more pronounced. By raising the significance level, we improve the accuracy with which the orthodox test identifies chains of higher order (from M_1 on up), but we deprove the accuracy with which it identifies random stationary sequences. Moreover, even with the significance level set as high as 20%, the Bayes test betters the orthodox test at M_3 and M_4 (and, *a fortiori*, at still higher order processes).

The irony is that Bayesian methods have been criticized from a minimax point of view. If the 'arbitrary' element of the Bayes method is guessed correctly (i.e., the prior), it can deliver excellent results. But if it is set incorrectly, or so it is claimed, the Bayes method performs disasterously. The very same criticism applies, *mutatis mutandis*, to the orthodox tests described in this chapter, where, now, the 'arbitrary' element is the significance level.

BIBLIOGRAPHY

Anderson, T. W. and Goodman, L. A.: 1957, 'Statistical Inference About Markov Chains', *Ann. Math. Stat.* **28**, 89–110.

Box, G. E. P. and G. Tiao: 1973, *Bayesian Inference in Statistical Analysis*, Addison–Wesley, Reading, Mass., Section 2.7.

Johnston, J.: 1972, *Econometrics* 2nd ed., McGraw-Hill, New York.

Kalbfleisch, J. G. and D. A. Sprott: 1970, 'Application of Likelihood Methods to Models Involving Large Numbers of Parameters', *J. Roy. Stat. Soc. B*, **32**, 175–194.

Lindley, D. V. and G. M. El-Sayyad: 1968, 'The Bayesian Estimation of a Linear Functional Relationship', *J. Roy. Stat. Soc., B* **30**, 190–202.

CHAPTER 12

COGNITIVE DECISIONS

Acceptance and rejection of hypotheses was considered from the perspective of error probabilities in Chapter 10. In this chapter, another approach, based on so-called 'epistemic utilities', will be canvassed.

1. EPIPHENOMENAL ACCEPTANCE

One may begin with the intuition that hypotheses ought to be accepted when their probabilities exceed a given threshold. Alternatively, hypotheses ought to be rejected when their probabilities fall below a critical threshold. But then given N equiprobable hypotheses (mutually exclusive and jointly exhaustive) and a critical threshold greater than $1/N$, all N hypotheses will be rejected even when one of them must be true (the so-called 'lottery paradox'). One way out, Hintikka's, is to couple the demand for high (or low) probability with the demand for much evidence. Given a partition of hypotheses, a sufficiently large sample will ordinarily give a highly peaked likelihood function; the probability of one hypothesis (the true one) will approach one, while the probabilities of the others tend to zero. Large samples also make it improbable that these comparative probabilities will be reversed at a later stage of sampling. Hintikka and Hiplinen (1970) have applied this approach to 'strong generalizations' (Chapter 4).

Wald's sequential analysis also reaches decisions of acceptance and rejection upon observing sufficiently large samples. As Szaniawski (1976) observes, the meaning of 'sufficiently large' can be fixed so as to insure: (i) prescribed error characteristics (essentially Wald's own approach), (ii) prescribed epistemic utility, or (iii) prescribed sample information. The three approaches, as Szaniawski, shows, are related in straightforward ways, given that the considered hypotheses have equal prior probabilities and given that all errors have equal disutility (compare Chapter 1, Section 5).

Only (iii) is relevant from a Bayesian standpoint: experimentation reflects a demand for information, and sampling ceases as soon as enough information has been obtained. More precisely, the scientist may be viewed as attaching different degrees of urgency to different questions (no doubt because answers to more urgent questions promise to throw more light on more urgent questions). The decision to discontinue sampling expresses a judgement that accumulating further evidence on the same point is of less urgency than equally costly information bearing on other questions. No suggestion that the hypothesis which happens to be ahead of the game when sampling ceases would necessarily continue to be ahead were sampling resumed is intended, much less any suggestion that it be held immune from revision.

We can, if we must, label these leading contenders 'accepted', but I cannot for the life of me see what is gained thereby. Many of those who have written on acceptance have been characteristically ambivalent on whether they mean more by 'accept H' than that H is 'ahead of the game' or that H has high probability based on much data. In much of the statistics literature, as we saw in Chapter 10, 'reject H' is just a *façon de parler* − a blind for 'regard the data as in poor agreement with H'. Other statisticians, like Neyman, mean no more by 'accept H' than that a certain action (optimal if H is true) be undertaken. In all of these approaches, acceptance and rejection are mere epiphenomena, and these inapposite terms could be easily dispensed with. Nevertheless, there are some (e.g., Giere − cf. Chapter 10, Section 4) who insist that hypotheses are 'really accepted' in science. It is those who hold that hypotheses are 'really accepted' *for broadly utilitarian reasons* whose views I wish to consider here.

2. ACCEPTANCE ACCORDING TO LEVI

It has been thought better to give a reasonably thorough discussion of one such position, Isaac Levi's, than to treat several superficially. Levi's recent retrospective article (Levi, 1976) throws much of his earlier thinking into sharper relief and indicates several important modifications of positions taken in his earlier book, *Gambling with Truth* (Levi, 1973). Unless otherwise indicated, references to Levi are to Levi (1976).

'Accepted' propositions are, for Levi, 'taken for granted' in subsequent inquiry; they are 'certain', 'infallible', and assigned probability one. They include the truths of logic and mathematics as a proper subset. Conversely, propositions which are 'rejected' in Levi's sense need not be contradictory or logically impossible. Rather, they are not 'serious' possibilities, and those who consider them otherwise (and do not assign them probability zero) are held to be neurotic or otherwise irrational (pp. 3−7). As an example, Levi considers the toss of a coin. The possibility that the coin lands on edge is 'serious', but the possibility that it will fly off to Alpha Centauri is not 'serious', and neither is the possibility that tossing it will cause the earth to explode. It isn't enough that we assign such possibilities miniscule probability; Levi insists that they have zero probability. For if the explosion of the earth consequent upon the flip of a coin is not assigned zero probability, Levi argues, it would be rational to refuse my offer to flip the coin and pay you $1000 whether it lands heads or tails. The disutility of having the earth blow up in your face presumably makes the expected utility of my offer negative, however small the positive probability you ascribe to the bizarre eventuality in question. To this Levi adds the further consideration that there are infinitely many such possibilities to consider in tossing a coin, and if all were assigned positive probability, the probability of their disjunction would be quite large and fail to reflect anybody's actual beliefs.

Levi's conclusion does not follow: we may assign 'unserious' possibilities infinitesimal probability. Infinitely many infinitesimals still add up to an infinitesimal, and an infinitesimal can combine with a very large (but finite) disutility to yield an infinitesimal, so that the expected utility is modest and realistic and scarcely affected by possible outcomes of infinitesimal probability. Quite apart from this consideration, it should be evident that Levi's first argument proves too much: I surely do not assign zero probability to being struck dead by a car when hurrying to make a light on my bicycle, but if Levi's argument was sound, I would never take such a risk. To defend what appears to be his equation of 'acting on' a proposition with assigning it probability one, Levi would have to show that all risky actions with possibly fatal consequences are irrational. But, I submit, anybody who was 'rational' in that sense would be, by ordinary standards, *irrationally* risk averse, indeed, neurotic.

Neither does premissing a proposition in science entail assigning it probability one. Indeed, many of the suppositions we premiss are believed strictly false, as when we premiss the randomness of a haphazardly drawn sample or the normality of a population. We exercise here what I. J. Good has labelled 'type II rationality', judging the errors incurred by these probably false assumptions as unlikely to invalidate our conclusions. Simplifying assumptions based on such sensitivity analysis is *de rigueur*, and we seldom give it a second thought. But to describe these assumptions as 'infallibly and certainly true' seems pretty clearly a misdescription.

That description must also be squared with Levi's claim that his 'infallibilism' is compatible with 'corrigibilism', with the claim that the bestowal of probability one (or probability zero) can subsequently be withdrawn. What could prompt such withdrawals? The answer, we learn (p. 30), is that

the hypothesis which prior to contraction was counted as certainly and necessarily false and after contraction is considered possibly true is informationally more attractive than the hypothesis removed from X's corpus in the contraction step and for this reason merits a hearing.

But then the demotion of 'infallible' propositions to a merely hypothetical status has nothing to do with the evidence bearing on their truth, but only with the erstwhile informational attractiveness of some alternative. I have no objection to the *retraction* (as opposed to the *revision* by conditionalization) of probabilities, when assumptions on which those probabilities were predicated are found to be in error. What I do object to is moving probabilities up or down (wilfully, as it were) because (p. 15)

we conclude that the trade-offs between risk of error and informational benefits are such as to warrant adding some hypothesis to the corpus and so to convert its status from mere hypothesis to settled, established and infallible truth.

This smacks of the decide-what-to-believe tradition criticized in Chapter 10 (cf. esp., Note 2).

There is, at the very least, a kind of equivocation Levi commits in passing from propositions which are accepted because their negations are not 'serious' to those which are accepted because the cognitive decision maker regards the risk of error thereby incurred as more than compensated by the informational benefits promised. Are we given to understand that all rational men would concur in that judgment? That would be assuming a lot, but if Levi does not assume it (which is my guess), then the two cases he treats are quite distinguishable, and our willingness to grant that some possibilities are not 'serious' connotes no willingness to grant that some propositions *ought* to be assigned probability one, not because their negations are not 'serious', given our evidence, but because their 'informational attractiveness' makes the risk of error incurred by 'adding them to one's corpus' worth running. So far as I can tell, the only argument Levi offers for this latter contention is the curiously perverse one (p. 13) that, by not accepting hypotheses in his strong sense, Bayesians run no risk of error.

Do scientists ever argue that a theory should be accounted infallibly true because its informational attractiveness makes the risk of so regarding it worth bearing? Not that this consideration is at all decisive, but I doubt it. For one thing, there are differences in the strength of the evidence supporting theories which might all be accepted on such grounds, and those differences are most naturally and most usefully reflected in different probability ascriptions. It is these differences ('evidential debits') and not the differing informational potentials *per se* which matter in gauging a novel theory which violates some more or less well entrenched assumption. The violation of the conservation of energy law by steady-state cosmology is an interesting case in point. Bondi (1960) writes:

Now, in fact, the mean density of the universe is so low, and the time scale of the universe so large, by comparison with terrestrial circumstances, that the process of continual creation required by the steady-state theory predicts the creation of only one hydrogen atom in the space the size of an ordinary living-room once every few million years. It is quite clear that this process, therefore, is in no way in conflict with the experiments on which the principle of the conservation of matter and energy is based. It is only in conflict with what was thought to be the simplest formulation of those experimental results namely, that matter and energy were precisely conserved. The steady-state theory has shown, however, that much simplicity can be gained in cosmology by the alternative formulation of a small amount of continual creation, with conservation beyond that. This may, therefore, be the formulation with the greatest overall simplicity.

I have chosen this case because, superficially considered, it may appear to exemplify Levi's analysis. Careful consideration, especially of the last sentence, shows, however, that it really does not. Bondi's point was not that the steady-state theory should be given a hearing because it had greater informational potential, but rather that a simpler overall theory (and, presumably, a better supported theory) would result by positing a steady-state universe (netting a considerable gain in simplicity) at the cost of having to posit the creation of a small amount of matter, thus sacrificing some simplicity. The argument is that the simplicity sacrificed in replacing exact by approximate conservation would be more than offset by the simplicity gained by assuming a steady-state universe.

Contrast Levi's treatment of simplicity or content. Content is a factor in the expected epistemic utility of accepting an hypothesis H (as strongest via induction) on the basis of an observation x, given by:

$$(12.1) \qquad U^*(H, x) = P^*H/x) - q\mathrm{Cont}(\overline{H}, x),$$

writing $\overline{H}$ for the negation of H. Here H may be any disjunction of elements of one's considered partition of hypotheses, q is an index of caution reflecting the experimenter's personal rate-of-exchange between the 'informational value' of H (or the relief from agnosticism which its acceptance affords) and the risk of error, and Cont is a measure of content.

The measure (12.1) is obtained from two mild assumptions: (i) that correct answers are epistemically preferred to errors, and (ii) that correct answers (resp. errors) of greater content are preferred (Levi, 1973, p. 76). *Levi's acceptance rule* then comes to this: first, reject an element of the partition for which (12.1) is negative, then accept as strongest via induction from x the disjunction H_x of all unrejected elements. If there are no other questions for which this rule accepts conflicting answers, then the 'locally' accepted hypothesis in question can be accepted in a stronger, 'global' sense which Levi labels *acceptance as evidence*. I will have nothing to say about this stronger notion, nor about the probabilities which enter (12.1), probabilities Levi appears to regard as subjective. I want to comment on the content measure and the caution index in turn.

In Levi (1973), content is measured by the logical improbability of an hypothesis, that is, by its improbability when all elements of the partition are assigned equal weight. Thus, if there are m basic hypotheses $H_1, \ldots, H_m$, $\text{Cont}(\overline{H_i}) = 1/m$, $i = 1, \ldots, m$. Similarly, if H is a disjunction of r of the m basic hypotheses, $\text{Cont}(\overline{H}) = r/m$. It follows that $\text{Cont}(H) = 1 - \text{Cont}(\overline{H})$. Levi's motivation is clear enough: the more alternative relevant answers an hypothesis excludes (among those not already excluded), the greater its content. The parenthetical qualifier is essential given the motivation, and this point cannot be overemphasized. Hence, $\text{Cont}(\overline{H}, x)$ is the content of $\overline{H}$ in the truncated partition induced by x (by omitting elements logically excluded by x). Thus, if x excludes n of m elements, $\text{Cont}(\overline{H_i}, x) = 0$ or $1/(m - n)$ according as H_i is among those excluded or not. In particular, $\text{Cont}(\overline{H_i}, x) = 1$ when x entails H_i.[1]

The apparatus of Levi (1973) was not intended to apply to theoretical hypotheses, for these will differ in other desirable traits, like simplicity and explanatory power. Utility functions for these other desiderata were to be employed and combined with the utility of truth and content to yield an overall composite epistemic utility function. The weights assigned different desiderata were admittedly arbitrary, and so too, for that matter, was the experimenter's specification of the caution index. Given this plethora of arbitrary and subjective elements, it is hard to see how the resulting decision theoretic formulation of theory choice "rationalizes the most important questions concerning the 'growth' or improvement of knowledge" rather than leaving them to be "psychologized, sociologized, or historicized" (p. 3). Indeed, in the absence of any clear indication what is to count as an 'epistemic utility' (or why), it is very hard to see how this formulation differs from methodological anarchy.

In the meantime, Levi has changed his formulation and now prefers to incorporate other desiderata or 'utilities' in the content function, at least when dealing with theoretical hypotheses. He writes (p. 40):

On this approach, considerations of simplicity, explanatory power, etc. no longer function as desiderata additional to informational value and truth which are to be averaged into the epistemic utility function. Rather they serve as factors which determine the informational values of rival hypotheses.

One would be tempted to think the modification a response to beckoning methodological anarchy but for the fact that it utterly fails to allay this fear and, in fact, the reason for the modification Levi cites is that it allows "treating all cases of inductive expansion in a unitary way" (p. 40). It is surely too much to expect different scientists to attach the same importance to different epistemic utilities. In that case, there will be no reason to expect them to choose the same measure of 'informational value', and methodological anarchy still reigns. Acceptance and rejection will depend, in an essential way, on what appear to be largely subjective or personal evaluations of the relative importance to be attached to a bewildering variety of proposed or proposable desiderata. In a real sense, each scientist becomes his own methodologist. Are we really compelled to follow Levi down this road?

An especially disagreeable feature of his present formulation is the dual role it assigns to simplicity and content. It seems to me at least as legitimate to relativize the content of an hypothesis or theory to the outcome space of the experiment (or experiments) which that hypothesis purports to explain as to relativize content to a considered partition of hypotheses. From the standpoint of accounting for experimental data, simplicity and content alike are plausibly measured by sample coverage (Chapter 5). At any rate, if an experimenter chose to assess 'informational value' in that way, Levi's rule of (local) acceptance (12.1) would, in effect, invite him to count content twice: once in his measure of 'informational value' and once in his probability measure (since sample coverage, we saw, influences the posterior probabilities through its influence on the average likelihoods). Once it is recognised that sample coverage (alias content, simplicity, specificity, overdetermination, etc.) is reflected in higher support and probability, the temptation to award theories of great simplicity an extra bonus on that account should be slight.

What are we to say about the caution index? In Levi (1967), p. 89, "the choice of q index" is said to be "a subjective factor, which does in some sense reflect the investigator's attitudes". If investigators are likely to differ in their choice of the caution index, the question immediately arises how consensus is to be attained. This is all the more so if they are also free to differ in their epistemic utility functions (i.e., in the weight they attach to possibly conflicting desiderata). These considerations may have prompted the following suggestion (Levi, 1960, Section IV):

... the canons of inference might require of each scientist that he have the same attitudes, assign the same utilities, or take each mistake with the same degree of seriousness as every other scientist.

This proposal was understandably dropped in subsequent publications of Levi's, but ambivalence over the status of the caution index resurfaces in Levi (1976). He there seems to suggest that some choices of q index might be excluded in some cases. To see why he thinks this, we must recall that his acceptance rule (Rule A of Levi (1967)) admits of iterative application (so-called 'bookkeeping'), as in the following example.

An urn contains three balls. Let H_i be the hypothesis that the number of black balls in the urn is i, $i = 0, 1, 2, 3$. Choose $q = 1$, and assume the hypotheses are initially equiprobable. A ball is drawn at random and proves to be black (outcome B). The posterior probabilities are 0, 1/6, 1/3, 1/2, resp., while $q\text{Cont}(-H_i, x) = 1/3$, $i = 1, 2, 3$, since H_0 is excluded by x. Hence, $H_x = H_2 \vee H_3$. If the conditions for global acceptance of H_x are met, it then becomes the new ultimate partition, and relative to this partition, the probabilities of H_2 and H_3 are 2/5 and 3/5, while $q\text{Cont}(-H_i, x)$ is now 1/2, $i = 2, 3$. H_2 is thus rejected on the next round, since $2/5 < 1/2$, and we are left with H_3. More generally, this iterative application of rule A with $q = 1$ will always reject all but the maximally likely hypothesis of the original partition. This is precisely the rule favored by Keith Lehrer (Lehrer (1976) and papers of his therein referred to). Levi rejects Lehrer's rule as "conflicting with presystematic judgments on one of those rare occasions where presystematic judgment is virtually noncontroversial" (p. 44). Namely, in flipping a fair coin 100 times, Lehrer's rule clearly rejects every hypothesis about the number of heads save H_{50}, and Levi balks at ruling out hypotheses asserting "that the relative frequency of heads will differ from 1/2 by some small amount". And to this he adds (p. 45):

I do not think that this constitutes an objection either to rule A, to bookkeeping, or to regarding the elements of the ultimate partition as equally informative. Rather it points to the dubiety of exercising the minimal degree of caution as indexed by $q = 1$.

At least as strong a case could be made for saying that the artificiality of acceptance *per se* is the source of the strain this result imposes on intuition. This is further argued by the fact that we obtain the same result in the coin problem with $q = 0.9$. Levi's exclusion of $q = 1$ makes no mention of the decision maker's actual demand for information, and it is, in this respect, reminiscent of Carnap's exclusion of the 'extreme methods' of his lambda continuum on a priori grounds (cf. Chapter 3, Section 6). If the agent's demand for information (relative to his risk averseness) is sufficiently great, why shouldn't he accept H_{50} as strongest? Is Levi claiming that no one's demands for information are ever that extreme? And if so, is this an empirical claim or a normative prescription?

Levi himself claims (pp. 45ff.) that $q = 1$ is appropriate in some cases. In estimating a statistical parameter θ with $q < 1$, we can never obtain more than an interval estimate. Yet, a particular value θ^* of the parameter (e.g., the value 0.5 of the linkage parameter) may be salient, and so the decision maker will want to alter his ultimate partition to test $\theta = \theta^*$ vs $\theta \neq \theta^*$, and 'throw caution to the winds' by setting $q = 1$. Notice, though, that the Bayes test (10.1) of Chapter 10 allows the data to register decisive support for the hypothesis $\theta = \theta^*$ without the need to adjust the parameters of Levi's system, viz., the caution index and the epistemic utility function. And the salient value of a statistical parameter is salient, I would add (and Levi tends to agree), not because it satisfies the demand for more information to a higher degree than other point estimates of the parameter, but because it can be assigned a lump of prior probability mass. E.g., the value 1/2 of the linkage parameter occurs when two genes lie on different chromosomes, and the a priori chance of this event is $(n - 1)/n$, where n is the number of chromosomes in a haploid set characteristic of the given species. Levi approaches the problem differently: one chooses a small value of δ and then accepts $\theta = \theta^*$ iff the interval estimate $(\theta^* - \delta, \theta^* + \delta)$ is accepted when $q = 1$. This has the effect of adding still another parameter, the length of the interval, to be manipulated by the investigator.

I can imagine circumstances in which one might want to estimate a statistical parameter with prescribed precision at a prescribed confidence level, whether for decisional purposes or to obtain a sensitive test of a theory. One would then calculate the smallest sample size at which such an estimate could be secured, and if the sampling costs were prohibitive, that would of course lead one to modify one's demands. A value of Levi's q determines the length of an accepted interval estimate, but, in practice, one would have to answer the question how precise an estimate is desired at a given confidence level before settling on an appropriate q. Like Hintikka's α (Chapter 4), Levi's q is not a readily interpretable parameter; one must decide how strong an hypothesis it would be sensible to 'accept' in the light of given data in order to set the value of q. This is already suggested by Levi's discussion of his own coin example. In the next section, I consider a closely related approach.

3. KUHN AND THE SOCIOLOGY OF SCIENCE

Critics have repeatedly charged Thomas Kuhn with denying that rational criteria guide theory choice. Certainly Kuhn's likening of 'paradigm shifts' to 'Gestalt switches' or 'conversion experiences', and his talk of scientific opponents as isolated in their own conceptual cages, doomed to argue past one another, has lent itself to this interpretation. So, too, has his evident sympathy with Max Planck's remark that new, revolutionary theories do not win the day because their opponents are convinced by the evidence, but because they die off leaving a younger generation, unfettered by the same theoretical blinders, and better able to appreciate the force of the arguments advanced on its behalf. Yet, for all that his rhetoric invites the charge that he is offering us 'mob psychology' in place of rational theory choice, Kuhn has steadfastly disavowed it, as in the following passage (Kuhn, 1970, p. 262):

What I am denying then is neither the existence of good reasons [for preferring one theory to another] nor that these reasons are of the sort usually described. I am, however, insisting that such reasons constitute values to be used in making choices rather than rules of choice. Scientists who share them may nevertheless make different choices in the same concrete situation. Two factors are deeply involved. First, in many concrete situations, different values, though all constitutive of good reasons, dictate different conclusions, different choices. In such cases of value conflict (e.g. one theory is simpler but the other is more accurate) the relative weight placed on different values by different individuals can play a decisive role in individual choice. More important, though scientists share these values and must continue to do so if science is to survive, they do not all apply them in the same way. Simplicity, scope, fruitfulness, and even accuracy can be judged quite differently (which is not to say they may be judged arbitrarily) by different people. Again, they may differ in their conclusions without violating any accepted rule.

That variability of judgement may ... even be essential to scientific advance. The choice of a research programme, involves major risks, particularly in its early stages. Some scientists must, by virtue of a value system differing in its applicability from the average, choose it early, or it will not be developed to the point of general persuasiveness. The choices dictated by these atypical value systems are, however, generally wrong. If all members of the community applied values in the same high-risk way, the group's enterprise would cease. This last point, I think, Lakatos misses, and with it the essential role of individual variability in what is only belatedly the unanimous decision of the group. As Feyerabend also emphasizes, to give these decisions a *'historical character'* or to suggest that they are made only *'with hindsight'* deprives them of their function. The scientific community cannot wait for history, though some individual members do. The needed results are instead achieved by distributing the risk that must be taken among the group's members.

Does anything in this argument suggest the appropriateness of phrases like decision by 'mob psychology'? I think not. On the contrary, one characteristic of a mob is its rejection of values which its members ordinarily share. Done by scientists, the result should be the end of their science, and the Lysenko case suggests that it would be. My argument, however, goes even further, for it emphasizes that, unlike most disciplines, the responsibility for applying shared scientific values, must be left to the specialists' group. It may not even be extended to all scientists, much less to all educated laymen, much less to the mob. If the specialists' group behaves like a mob, renouncing its normal values, then science is already past saving.

The picture that emerges here is sociological: there are many hallmarks of a good theory (simplicity, scope, fruitfulness, and accuracy are mentioned), but they are objective only in the sense that a scientific community recognizes their importance. The different practitioners of a discipline will interpret them variously, and of course, will attach different importance to possibly conflicting desiderata, like simplicity and accuracy. This conception, which is close to Levi's in the role it assigns to 'epistemic utilities' and risk, stops short of methodological anarchy only in its suggestion that there are (albeit, imprecise) limits on the range of interpretations of the different desiderata which a scientific community will countenance. This reply of Kuhn's to the charge of 'mob psychology' will seem lame to many who press this charge.

He attempts to buttress the reply by likening scientific progress to evolutionary progress. Within any scientific community, there will be a dominant group (or, perhaps, in pre-paradigm stages, several schools or centers of influence) characterized by a floating body of shared beliefs and methodological biases. But there will also be a small number of sports who diverge from this received 'dogma' (Kuhn's term) in varying degrees. (Good sociological reasons for their existence will not be far to seek.) The prevailing diversity of viewpoints will insure that all but the most outlandish theories (or germinal ideas for theories) get a hearing. Nevertheless, those which fail to attain reliably good fit to a reasonably well defined class of experiments despite the best efforts of their proponents to refine them (short of rendering them hopelessly complicated or nearly vacuous) will be 'selected against' (or, rather, their proponents will be 'selected against' and eventually shrink in number). The (collective) rationality of science therefore neither exhibits nor requires complete methodological agreement, much less strict adherence to methodological rules.

Kuhn has attempted to steer a passage between the extremes of methodological dogmatism and methodological anarchism; I am strongly in sympathy with this attempt. Moreover, Kuhn's evolutionary account of theory change is not, at any rate, a grossly inaccurate description of what actually transpires. But neither is it entirely accurate. Scientific disputes are often recalcitrant, but that in itself does not argue the correctness of Kuhn's portrayal. A strong case can be made for the primacy of evidence, and recalcitrant disagreement can be accounted for by conflicting data as well as by methodological divergence. The immediate successors of Copernicus had to balance the greater support that accrued to the heliostatic theory from planetary data as the result of its greater simplicity (ch. 7) against physical objections to the earth's motion. It is not easy to see how an overall assessment of support could be arrived at for this case without arbitrarily weighting the different (celestial and terrestrial) strands of evidence. Kuhn's evolutionary picture continues to seem apt here, not, however, because different scientists weight the various desiderata differently, but because they diverge in the importance they attach to different aspects of the data.

Let us leave the descriptive plane for a moment and turn to the normative. From the Bayesian standpoint advanced here, probability (or support) is the single yardstick by which theories should be compared or evaluated. The desiderata which are normally cited are genuine and important only insofar as they are reflected in support. That already suggests a drastic narrowing of the scope of methodological divergence.

I would even press the evolutionary analogy farther and argue that adaptation occurs at the methodological level: scientists gradually become better attuned to those features of theories that render them better confirmed by conforming data. The inferential task of the scientist has always been viewed as that of fitting the simplest possible (or simplest plausible) theory to the extant data. By sharpening notions like simplicity in ways that reveal their connection with support, the methodologist can help circumvent much inconclusive discussion generated by appeals to dubious merits of theories or still more dubious methodological fiats. The history of genetics, to take a single example,[2] is littered with misuses of Occam's Razor, that is, uses based on vague or irrelevant senses of 'simplicity'. (Many scientists are rightly sceptical of such appeals.) At the same time, by laying bare these connections, methodology may hope to accelerate the adaptation to which I allude and thereby render a contribution to heuristics.

Finally, by invoking the yardstick of support, a great many apparently arbitrary weightings of conflicting desiderata are eliminated, yet further reducing the scope for legitimate disagreement. Kuhn mentions the alleged need to weight simplicity and accuracy where the simpler theory of a comparison is also the less accurate. The weighting is accomplished, however, in a mathematically determinate and non-judgmental way by the use of average likelihood (Chapter 5). I have emphasized throughout the extent to which Bayesian methods determine weightings or trade-offs that are alleged to be, at best, judgmental, and at worst, completely arbitrary.

Even current scientific practice multiplies opportunity for disagreement beyond necessity. It is common practice to compare theories, for example, by comparing their predictions over a battery of statistics. Thus, two learning theories might be compared by comparing their predicted mean learning curves, total number of errors prior to learning, variance of total errors, and other statistics obtained by averaging over subjects.[4] This practice is objectionable on several counts. Why choose those particular statistics? Are simplicity differences reflected? Finally, unless one theory's predictions are uniformly better than the other's, opinions will differ over which of the two is better supported by the data. All of these difficulties are averted by computing average likelihoods from the raw data (in the learning example, from the actual subject-item sequences).

4. CONCLUSION

The objectivist Bayesian position offers an appealing 'middle way'. It disavows attempts to straitjacket the scientist by prescribing rules of acceptance and rejection. It does not, for example, offer directives when to abandon (rather than go on complicating) a theory; rather, it provides indices of probability or support, and these, I have maintained, are sufficient to direct the course of research along fruitful lines and secure the needed consensus. At the same time, the objectivist Bayesian position avoids the extremes of subjectivism and anarchism. By entering elements others leave to personal judgment explicitly into the analysis, greater objectivity becomes attainable. More is to be gained by refining these indices of support and widening their scope than by laying down arbitrary precepts with no clear or compelling rationale or by abandoning the field to the historians, sociologists and psychologists of science.

NOTES

[1] Levi (1971), pp. 870–871, rejects Lehrer's acceptance rule (cf. below), which rejects all but the maximally likely hypothesis, because it violates his requirement that the more

informative of two erroneous hypotheses be accorded higher epistemic utility, whereas Lehrer, Jeffrey and others maintain that one error is no worse than another. (Another objection of Levi's to Lehrer's rule is considered below.) Levi offers no real argument on behalf of this requirement, only more 'circumlocuitous reaffirmation'. My own intuitions on the matter (insofar as I would admit to having intuitions about acceptance) are wholly unformed. My suspicion is that the disutility attached to accepting an erroneous hypothesis will depend greatly on the specific problem context. Certainly Levi makes no case for saying that 'the proximate goals of inquiry' universally dictate that more of a bad thing (erroneous information) be preferred to less. It is sobering to think that differences this diaphanous divide Levi from Lehrer. By confining attention to Levi's rule, this chapter gives little hint of the rich diversity of acceptance rules that have been proposed, a diversity that further argues the lack of any very clear concept of acceptance.

[2] Let x_i entail H_i. Then Cont$(-H_j, x_i)$ and $P(H_j/x_i)$ are both 1 or 0 according as $i = j$ or not. Thus, when $q = 1$, Levi's rule A rejects no H_j in the light of a *perfectly definitive outcome* x_i. This difficulty suggests omitting from H_x all H_j for which $P(H_j/x_i) \leq q$Cont$(-H_j, x_i)$. This amounts to changing Levi's rule for breaking ties, opting for content where he opts for safety. But the revised rule runs afoul of outcomes x which induce a uniform posterior distribution of the H_j, for then every 'ultimate' hypothesis H_j is rejected. Better, perhaps, to stick with the original rule A and stipulate that logically excluded hypotheses be rejected. (Perhaps this is what Levi intends?) I shall assume this emendation in what follows. There is, however, another related difficulty. As Lehrer (1969) shows, an H_j which is accepted may fail to be accepted when probabilities are altered in its favor (i.e., a disjunction including it may then be accepted as strongest). This can even happen where that H_j widens its margin over its nearest rival, and it can even happen if bookkeeping (cf. below) is allowed, as the following example (a slight modification of Lehrer's) shows. In case one, we are given 3 red counters, 3 white counters, and 9 blue counters. A counter is drawn at random, and the three hypotheses about the outcome are R(red), W(white), and B(blue). Let $q = 0.61$. Then $H_x = B$. Now suppose that two of the red counters and one blue counter are omitted, giving 1 red, 3 white, and 8 blue. In this case, $H_x = W \vee B$, even though the probability (and margin) of B has increased, and the result is unchanged if we reapply rule A to the reduced partition H_x (bookkeeping). Levi (1971), p. 868, dismisses this objection by claiming it presupposes that acceptance rules should depend on error probabilities and probability differences. Certainly the condition Lehrer proposes is a prima facie reasonable one: that an hypothesis accepted as strongest should continue to be accepted as strongest when the odds in its favor and its margin both increase. But, here again, the lack of any very clear concept of acceptance makes it impossible to adjudicate the issue.

[3] As discussed by Carlson (1966), Chapter 3–12, 26, 27.

[4] Cf. Atkinson *et al.* (1965), Chapter 3.

BIBLIOGRAPHY

Atkinson, R. C. *et al.*: 1965, *Introduction to Mathematical Learning Theory*, Wiley, New York.

Bogdan, R. (ed.): 1976, *Local Induction*, D. Reidel, Dordrecht–Holland.

Bondi, H. *et al.*: 1960, *Rival Theories of Cosmology*, Oxford.

Cardwell, C.: 1971, 'Gambling for Content', *J. Phil.* **LXVIII**, 860–864.

Carlson, E. A.: 1966, *The Gene: A Critical History*, W. B. Saunders, Philadelphia.

Goossens, W. K.: 'A Critique of Epistemic Utilities', in Bogdan (1976), pp. 93–114.

Hacking, I.: 1967, Review of Levi, 'Gambling with Truth', *Synthese* **17**, 444–448.

Hilpinen, R.: 1968, *Rules of Acceptance and Inductive Logic*, North-Holland, Amsterdam.

Hintikka, K. J. J. and Hilpinen, R.: 1970, 'Knowledge, Acceptance, and Inductive Logic', *Information and Inference* (K. J. J. Hintikka and P. Suppes, eds.), D. Reidel, Dordrecht–Holland, pp. 1–20.

Jeffrey, R.: 1968, Review of Levi, 'Gambling with Truth', *J. Phil.* **65**, 313–322.

Kuhn, T.: 1970, 'Reflections on My Critics', *Criticism and the Growth of Knowledge* (I. Lakatos and A. Musgrave, eds.), Cambridge.

Lehrer, K.: 1969, 'Induction: a Consistent Gamble', *Nous* **3**, 285–297.

Lehrer, K.: 1976, 'Induction, Consensus, and Catastrophe', in Bogdan (1976).

Levi, I.: 1960, 'Must the Scientist Make Value Judgments?', *J. Phil.* **57**, 345–357.

Levi, I.: 1971, 'Truth, Content, and Ties', *J. Phil.* **68**, 865–876.

Levi, I.: 1973, *Gambling with Truth*, 2nd ed., MIT Press, Cambridge, Mass.

Levi, I.: 1976, 'Acceptance Revisited', in Bogdan (1976), pp. 1–71.

Szaniawski, K.: 1976, 'On Sequential Inference', in Bogdan (1976), pp. 171–182.

SUBJECT INDEX

AUTHOR INDEX

SYNTHESE LIBRARY

Monographs on Epistemology, Logic, Methodology,
Philosophy of Science, Sociology of Science and of Knowledge, and on the
Mathematical Methods of Social and Behavioral Sciences

Managing Editor:
JAAKKO HINTIKKA (Academy of Finland and Stanford University)

Editors:

ROBERT S. COHEN (Boston University)
DONALD DAVIDSON (University of Chicago)
GABRIËL NUCHELMANS (University of Leyden)
WESLEY C. SALMON (University of Arizona)

1. J. M. Bocheński, *A Precis of Mathematical Logic.* 1959, X + 100 pp.
2. P. L. Guiraud, *Problèmes et méthodes de la statistique linguistique.* 1960, VI + 146 pp.
3. Hans Freudenthal (ed.), *The Concept and the Role of the Model in Mathematics and Natural and Social Sciences, Proceedings of a Colloquium held at Utrecht, The Netherlands, January 1960.* 1961, VI + 194 pp.
4. Evert W. Beth, *Formal Methods. An Introduction to Symbolic Logic and the Study of Effective Operations in Arithmetic and Logic.* 1962, XIV + 170 pp.
5. B. H. Kazemier and D. Vuysje (eds.), *Logic and Language. Studies Dedicated to Professor Rudolf Carnap on the Occasion of His Seventieth Birthday.* 1962, VI + 256 pp.
6. Marx W. Wartofsky (ed.), *Proceedings of the Boston Colloquium for the Philosophy of Science, 1961-1962,* Boston Studies in the Philosophy of Science (ed. by Robert S. Cohen and Marx W. Wartofsky), Volume I. 1973, VIII + 212 pp.
7. A. A. Zinov'ev, *Philosophical Problems of Many-Valued Logic.* 1963, XIV + 155 pp.
8. Georges Gurvitch, *The Spectrum of Social Time.* 1964, XXVI + 152 pp.
9. Paul Lorenzen, *Formal Logic.* 1965, VIII + 123 pp.
10. Robert S. Cohen and Marx W. Wartofsky (eds.), *In Honor of Philipp Frank,* Boston Studies in the Philosophy of Science (ed. by Robert S. Cohen and Marx W. Wartofsky), Volume II. 1965, XXXIV + 475 pp.
11. Evert W. Beth, *Mathematical Thought. An Introduction to the Philosophy of Mathematics.* 1965, XII + 208 pp.
12. Evert W. Beth and Jean Piaget, *Mathematical Epistemology and Psychology.* 1966, XII + 326 pp.
13. Guido Küng, *Ontology and the Logistic Analysis of Language. An Enquiry into the Contemporary Views on Universals.* 1967, XI + 210 pp.
14. Robert S. Cohen and Marx W. Wartofsky (eds.), *Proceedings of the Boston Colloquium for the Philosophy of Science 1964-1966, in Memory of Norwood Russell Hanson,* Boston Studies in the Philosophy of Science (ed. by Robert S. Cohen and Marx W. Wartofsky), Volume III. 1967, XLIX + 489 pp.

15. C. D. Broad, *Induction, Probability, and Causation. Selected Papers.* 1968, XI + 296 pp.
16. Günther Patzig, *Aristotle's Theory of the Syllogism. A Logical-Philosophical Study of Book A of the Prior Analytics.* 1968, XVII + 215 pp.
17. Nicholas Rescher, *Topics in Philosophical Logic.* 1968, XIV + 347 pp.
18. Robert S. Cohen and Marx W. Wartofsky (eds.), *Proceedings of the Boston Colloquium for the Philosophy of Science 1966-1968,* Boston Studies in the Philosophy of Science (ed. by Robert S. Cohen and Marx W. Wartofsky), Volume IV. 1969, VIII + 537 pp.
19. Robert S. Cohen and Marx W. Wartofsky (eds.), *Proceedings of the Boston Colloquium for the Philosophy of Science 1966-1968,* Boston Studies in the Philosophy of Science (ed. by Robert S. Cohen and Marx W. Wartofsky), Volume V. 1969, VIII + 482 pp.
20. J.W. Davis, D. J. Hockney, and W. K. Wilson (eds.), *Philosophical Logic.* 1969, VIII + 277 pp.
21. D. Davidson and J. Hintikka (eds.), *Words and Objections: Essays on the Work of W.V. Quine.* 1969, VIII + 366 pp.
22. Patrick Suppes, *Studies in the Methodology and Foundations of Science. Selected Papers from 1911 to 1969.* 1969, XII + 473 pp.
23. Jaakko Hintikka, *Models for Modalities. Selected Essays.* 1969, IX + 220 pp.
24. Nicholas Rescher *et al.* (eds.), *Essays in Honor of Carl G. Hempel. A Tribute on the Occasion of His Sixty-Fifth Birthday.* 1969, VII + 272 pp.
25. P. V. Tavanec (ed.), *Problems of the Logic of Scientific Knowledge.* 1969, XII + 429 pp.
26. Marshall Swain (ed.), *Induction, Acceptance, and Rational Belief.* 1970, VII + 232 pp.
27. Robert S. Cohen and Raymond J. Seeger (eds.), *Ernst Mach: Physicist and Philosopher,* Boston Studies in the Philosophy of Science (ed. by Robert S. Cohen and Marx W. Wartofsky), Volume VI. 1970, VIII + 295 pp.
28. Jaakko Hintikka and Patrick Suppes, *Information and Inference.* 1970, X + 336 pp.
29. Karel Lambert, *Philosophical Problems in Logic. Some Recent Developments.* 1970, VII + 176 pp.
30. Rolf A. Eberle, *Nominalistic Systems.* 1970, IX + 217 pp.
31. Paul Weingartner and Gerhard Zecha (eds.), *Induction, Physics, and Ethics: Proceedings and Discussions of the 1968 Salzburg Colloquium in the Philosophy of Science.* 1970, X + 382 pp.
32. Evert W. Beth, *Aspects of Modern Logic.* 1970, XI + 176 pp.
33. Risto Hilpinen (ed.), *Deontic Logic: Introductory and Systematic Readings.* 1971, VII + 182 pp.
34. Jean-Louis Krivine, *Introduction to Axiomatic Set Theory.* 1971, VII + 98 pp.
35. Joseph D. Sneed, *The Logical Structure of Mathematical Physics.* 1971, XV + 311 pp.
36. Carl R. Kordig, *The Justification of Scientific Change.* 1971, XIV + 119 pp.
37. Milič Čapek, *Bergson and Modern Physics,* Boston Studies in the Philosophy of Science (ed. by Robert S. Cohen and Marx W. Wartofsky), Volume VII. 1971, XV + 414 pp.

38. Norwood Russell Hanson, *What I Do Not Believe, and Other Essays* (ed. by Stephen Toulmin and Harry Woolf), 1971, XII + 390 pp.
39. Roger C. Buck and Robert S. Cohen (eds.), *PSA 1970. In Memory of Rudolf Carnap*, Boston Studies in the Philosophy of Science (ed. by Robert S. Cohen and Marx W. Wartofsky), Volume VIII. 1971, LXVI + 615 pp. Also available as paperback.
40. Donald Davidson and Gilbert Harman (eds.), *Semantics of Natural Language.* 1972, X + 769 pp. Also available as paperback.
41. Yehoshua Bar-Hillel (ed.), *Pragmatics of Natural Languages.* 1971, VII + 231 pp.
42. Sören Stenlund, *Combinators, λ-Terms and Proof Theory.* 1972, 184 pp.
43. Martin Strauss, *Modern Physics and Its Philosophy. Selected Papers in the Logic, History, and Philosophy of Science.* 1972, X + 297 pp.
44. Mario Bunge, *Method, Model and Matter.* 1973, VII + 196 pp.
45. Mario Bunge, *Philosophy of Physics.* 1973, IX + 248 pp.
46. A. A. Zinov'ev, *Foundations of the Logical Theory of Scientific Knowledge (Complex Logic)*, Boston Studies in the Philosophy of Science (ed. by Robert S. Cohen and Marx W. Wartofsky), Volume IX. Revised and enlarged English edition with an appendix, by G. A. Smirnov, E. A. Sidorenka, A. M. Fedina, and L. A. Bobrova. 1973, XXII + 301 pp. Also available as paperback.
47. Ladislav Tondl, *Scientific Procedures*, Boston Studies in the Philosophy of Science (ed. by Robert S. Cohen and Marx W. Wartofsky), Volume X. 1973, XII + 268 pp. Also available as paperback.
48. Norwood Russell Hanson, *Constellations and Conjectures* (ed. by Willard C. Humphreys, Jr.). 1973, X + 282 pp.
49. K. J. J. Hintikka, J. M. E. Moravcsik, and P. Suppes (eds.), *Approaches to Natural Language. Proceedings of the 1970 Stanford Workshop on Grammar and Semantics.* 1973, VIII + 526 pp. Also available as paperback.
50. Mario Bunge (ed.), *Exact Philosophy – Problems, Tools, and Goals.* 1973, X + 214 pp.
51. Radu J. Bogdan and Ilkka Niiniluoto (eds.), *Logic, Language, and Probability. A Selection of Papers Contributed to Sections IV, VI, and XI of the Fourth International Congress for Logic, Methodology, and Philosophy of Science, Bucharest, September 1971.* 1973, X + 323 pp.
52. Glenn Pearce and Patrick Maynard (eds.), *Conceptual Chance.* 1973, XII + 282 pp.
53. Ilkka Niiniluoto and Raimo Tuomela, *Theoretical Concepts and Hypothetico-Inductive Inference.* 1973, VII + 264 pp.
54. Roland Fraïssé, *Course of Mathematical Logic – Volume 1: Relation and Logical Formula.* 1973, XVI + 186 pp. Also available as paperback.
55. Adolf Grünbaum, *Philosophical Problems of Space and Time.* Second, enlarged edition, Boston Studies in the Philosophy of Science (ed. by Robert S. Cohen and Marx W. Wartofsky), Volume XII. 1973, XXIII + 884 pp. Also available as paperback.
56. Patrick Suppes (ed.), *Space, Time, and Geometry.* 1973, XI + 424 pp.
57. Hans Kelsen, *Essays in Legal and Moral Philosophy*, selected and introduced by Ota Weinberger. 1973, XXVIII + 300 pp.
58. R. J. Seeger and Robert S. Cohen (eds.), *Philosophical Foundations of Science. Proceedings of an AAAS Program, 1969*, Boston Studies in the Philosophy of

Science (ed. by Robert S. Cohen and Marx W. Wartofsky), Volume XI. 1974, X + 545 pp. Also available as paperback.

59. Robert S. Cohen and Marx W. Wartofsky (eds.), *Logical and Epistemological Studies in Contemporary Physics*, Boston Studies in the Philosophy of Science (ed. by Robert S. Cohen and Marx W. Wartofsky), Volume XIII. 1973, VIII + 462 pp. Also available as paperback.

60. Robert S. Cohen and Marx W. Wartofsky (eds.), *Methodological and Historical Essays in the Natural and Social Sciences. Proceedings of the Boston Colloquium for the Philosophy of Science, 1969-1972*, Boston Studies in the Philosophy of Science (ed. by Robert S. Cohen and Marx W. Wartofsky), Volume XIV. 1974, VIII + 405 pp. Also available as paperback.

61. Robert S. Cohen, J. J. Stachel and Marx W. Wartofsky (eds.), *For Dirk Struik. Scientific, Historical and Political Essays in Honor of Dirk J. Struik*, Boston Studies in the Philosophy of Science (ed. by Robert S. Cohen and Marx W. Wartofsky), Volume XV. 1974, XXVII + 652 pp. Also available as paperback.

62. Kazimierz Ajdukiewicz, *Pragmatic Logic*, transl. from the Polish by Olgierd Wojtasiewicz. 1974, XV + 460 pp.

63. Sören Stenlund (ed.), *Logical Theory and Semantic Analysis. Essays Dedicated to Stig Kanger on His Fiftieth Birthday*. 1974, V + 217 pp.

64. Kenneth F. Schaffner and Robert S. Cohen (eds.), *Proceedings of the 1972 Biennial Meeting, Philosophy of Science Association*, Boston Studies in the Philosophy of Science (ed. by Robert S. Cohen and Marx W. Wartofsky), Volume XX. 1974, IX + 444 pp. Also available as paperback.

65. Henry E. Kyburg, Jr., *The Logical Foundations of Statistical Inference*. 1974, IX + 421 pp.

66. Marjorie Grene, *The Understanding of Nature: Essays in the Philosophy of Biology*, Boston Studies in the Philosophy of Science (ed. by Robert S. Cohen and Marx W. Wartofsky), Volume XXIII. 1974, XII + 360 pp. Also available as paperback.

67. Jan M. Broekman, *Structuralism: Moscow, Prague, Paris*. 1974, IX + 117 pp.

68. Norman Geschwind, *Selected Papers on Language and the Brain*, Boston Studies in the Philosophy of Science (ed. by Robert S. Cohen and Marx W. Wartofsky), Volume XVI. 1974, XII + 549 pp. Also available as paperback.

69. Roland Fraïssé, *Course of Mathematical Logic* – Volume 2: *Model Theory*. 1974, XIX + 192 pp.

70. Andrzej Grzegorczyk, *An Outline of Mathematical Logic. Fundamental Results and Notions Explained with All Details*. 1974, X + 596 pp.

71. Franz von Kutschera, *Philosophy of Language*. 1975, VII + 305 pp.

72. Juha Manninen and Raimo Tuomela (eds.), *Essays on Explanation and Understanding. Studies in the Foundations of Humanities and Social Sciences*. 1976, VII + 440 pp.

73. Jaakko Hintikka (ed.), *Rudolf Carnap, Logical Empiricist. Materials and Perspectives*. 1975, LXVIII + 400 pp.

74. Milič Čapek (ed.), *The Concepts of Space and Time. Their Structure and Their Development*, Boston Studies in the Philosophy of Science (ed. by Robert S. Cohen and Marx W. Wartofsky), Volume XXII. 1976, LVI + 570 pp. Also available as paperback.

75. Jaakko Hintikka and Unto Remes, *The Method of Analysis. Its Geometrical Origin and Its General Significance,* Boston Studies in the Philosophy of Science (ed. by Robert S. Cohen and Marx W. Wartofsky), Volume XXV. 1974, XVIII + 144 pp. Also available as paperback.

76. John Emery Murdoch and Edith Dudley Sylla, *The Cultural Context of Medieval Learning. Proceedings of the First International Colloquium on Philosophy, Science, and Theology in the Middle Ages − September 1973,* Boston Studies in the Philosophy of Science (ed. by Robert S. Cohen and Marx W. Wartofsky), Volume XXVI. 1975, X + 566 pp. Also available as paperback.

77. Stefan Amsterdamski, *Between Experience and Metaphysics. Philosophical Problems of the Evolution of Science,* Boston Studies in the Philosophy of Science (ed. by Robert S. Cohen and Marx W. Wartofsky), Volume XXXV. 1975, XVIII + 193 pp. Also available as paperback.

78. Patrick Suppes (ed.), *Logic and Probability in Quantum Mechanics.* 1976, XV + 541 pp.

79. H. von Helmholtz, *Epistemological Writings.* (A New Selection Based upon the 1921 Volume edited by Paul Hertz and Moritz Schlick, Newly Translated and Edited by R. S. Cohen and Y. Elkana), Boston Studies in the Philosophy of Science, Volume XXXVII. 1977 (forthcoming).

80. Joseph Agassi, *Science in Flux,* Boston Studies in the Philosophy of Science (ed. by Robert S. Cohen and Marx W. Wartofsky), Volume XXVIII. 1975, XXVI + 553 pp. Also available as paperback.

81. Sandra G. Harding (ed.), *Can Theories Be Refuted? Essays on the Duhem-Quine Thesis.* 1976, XXI + 318 pp. Also available as paperback.

82. Stefan Nowak, *Methodology of Sociological Research: General Problems.* 1977, XVIII + 504 pp. (forthcoming).

83. Jean Piaget, Jean-Blaise Grize, Alina Szeminska, and Vinh Bang, *Epistemology and Psychology of Functions.* 1977 (forthcoming).

84. Marjorie Grene and Everett Mendelsohn (eds.), *Topics in the Philosophy of Biology,* Boston Studies in the Philosophy of Science (ed. by Robert S. Cohen and Marx W. Wartofsky), Volume XXVII. 1976, XIII + 454 pp. Also available as paperback.

85. E. Fischbein, *The Intuitive Sources of Probabilistic Thinking in Children.* 1975, XIII + 204 pp.

86. Ernest W. Adams, *The Logic of Conditionals. An Application of Probability to Deductive Logic.* 1975, XIII + 156 pp.

87. Marian Przełęcki and Ryszard Wójcicki (eds.), *Twenty-Five Years of Logical Methodology in Poland.* 1977, VIII + 803 pp. (forthcoming).

88. J. Topolski, *The Methodology of History.* 1976, X + 673 pp.

89. A. Kasher (ed.), *Language in Focus: Foundations, Methods and Systems. Essays Dedicated to Yehoshua Bar-Hillel,* Boston Studies in the Philosophy of Science (ed. by Robert S. Cohen and Marx W. Wartofsky), Volume XLIII. 1976, XXVIII + 679 pp. Also available as paperback.

90. Jaakko Hintikka, *The Intentions of Intentionality and Other New Models for Modalities.* 1975, XVIII + 262 pp. Also available as paperback.

91. Wolfgang Stegmüller, *Collected Papers on Epistemology, Philosophy of Science and History of Philosophy,* 2 Volumes, 1977 (forthcoming).

92. Dov M. Gabbay, *Investigations in Modal and Tense Logics with Applications to Problems in Philosophy and Linguistics*. 1976, XI + 306 pp.

93. Radu J. Bogdan, *Local Induction*. 1976, XIV + 340 pp.

94. Stefan Nowak, *Understanding and Prediction: Essays in the Methodology of Social and Behavioral Theories*. 1976, XIX + 482 pp.

95. Peter Mittelstaedt, *Philosophical Problems of Modern Physics*, Boston Studies in the Philosophy of Science (ed. by Robert S. Cohen and Marx W. Wartofsky), Volume XVIII. 1976, X + 211 pp. Also available as paperback.

96. Gerald Holton and William Blanpied (eds.), *Science and Its Public: The Changing Relationship*, Boston Studies in the Philosophy of Science (ed. by Robert S. Cohen and Marx W. Wartofsky), Volume XXXIII. 1976, XXV + 289 pp. Also available as paperback.

97. Myles Brand and Douglas Walton (eds.), *Action Theory. Proceedings of the Winnipeg Conference on Human Action, Held at Winnipeg, Manitoba, Canada, 9-11 May 1975*. 1976, VI + 345 pp.

98. Risto Hilpinen, *Knowledge and Rational Belief*. 1978 (forthcoming).

99. R. S. Cohen, P. K. Feyerabend, and M. W. Wartofsky (eds.), *Essays in Memory of Imre Lakatos*, Boston Studies in the Philosophy of Science (ed. by Robert S. Cohen and Marx W. Wartofsky), Volume XXXIX. 1976, XI + 762 pp. Also available as paperback.

100. R. S. Cohen and J. Stachel (eds.), *Leon Rosenfeld, Selected Papers*. Boston Studies in the Philosophy of Science (ed. by Robert S. Cohen and Marx W. Wartofsky), Volume XXI. 1977 (forthcoming).

101. R. S. Cohen, C. A. Hooker, A. C. Michalos, and J. W. van Evra (eds.), *PSA 1974: Proceedings of the 1974 Biennial Meeting of the Philosophy of Science Association*, Boston Studies in the Philosophy of Science (ed. by Robert S. Cohen and Marx W. Wartofsky), Volume XXXII. 1976, XIII + 734 pp. Also available as paperback.

102. Yehuda Fried and Joseph Agassi, *Paranoia: A Study in Diagnosis*, Boston Studies in the Philosophy of Science (ed. by Robert S. Cohen and Marx W. Wartofsky), Volume L. 1976, XV + 212 pp. Also available as paperback.

103. Marian Przełęcki, Klemens Szaniawski, and Ryszard Wójcicki (eds.), *Formal Methods in the Methodology of Empirical Sciences*. 1976, 455 pp.

104. John M. Vickers, *Belief and Probability*. 1976, VIII + 202 pp.

105. Kurt H. Wolff, *Surrender and Catch: Experience and Inquiry Today*, Boston Studies in the Philosophy of Science (ed. by Robert S. Cohen and Marx W. Wartofsky), Volume LI. 1976, XII + 410 pp. Also available as paperback.

106. Karel Kosík, *Dialectics of the Concrete*, Boston Studies in the Philosophy of Science (ed. by Robert S. Cohen and Marx W. Wartofsky), Volume LII. 1976, VIII + 158 pp. Also available as paperback.

107. Nelson Goodman, *The Structure of Appearance*, Boston Studies in the Philosophy of Science (ed. by Robert S. Cohen and Marx W. Wartofsky), Volume LIII. 1977 (forthcoming).

108. Jerzy Giedymin (ed.), *Kazimierz Ajdukiewicz: Scientific World-Perspective and Other Essays, 1931–1963*. 1977 (forthcoming).

109. Robert L. Causey, *Unity of Science*. 1977, VIII+185 pp.

110. Richard Grandy, *Advanced Logic for Applications*. 1977 (forthcoming).

111. Robert P. McArthur, *Tense Logic*. 1976, VII + 84 pp.
112. Lars Lindahl, *Position and Change: A Study in Law and Logic*. 1977, IX + 299 pp.
113. Raimo Tuomela, *Dispositions*. 1977 (forthcoming).
114. Herbert A. Simon, *Models of Discovery and Other Topics in the Methods of Science*, Boston Studies in the Philosophy of Science (ed. by Robert S. Cohen and Marx W. Wartofsky), Volume LIV. 1977 (forthcoming).
115. Roger D. Rosenkrantz, *Inference, Method and Decision*. 1977 (forthcoming).
116. Raimo Tuomela, *Human Action and Its Explanation. A Study on the Philosophical Foundations of Psychology*. 1977 (forthcoming).
117. Morris Lazerowitz, *The Language of Philosophy*, Boston Studies in the Philosophy of Science (ed. by Robert S. Cohen and Marx W. Wartofsky), Volume LV. 1977 (forthcoming).
118. Tran Duc Thao, *Origins of Language and Consciousness*, Boston Studies in the Philosophy of Science (ed. by Robert S. Cohen and Marx. W. Wartofsky), Volume LVI. 1977 (forthcoming).
119. Jerzy Pelc, *Polish Semiotic Studies, 1894–1969*. 1977 (forthcoming).
120. Ingmar Pörn, *Action Theory and Social Science. Some Formal Models*. 1977 (forthcoming).
121. Joseph Margolis, *Persons and Minds*, Boston Studies in the Philosophy of Science (ed. by Robert S. Cohen and Marx W. Wartofsky), Volume LVII. 1977 (forthcoming).

SYNTHESE HISTORICAL LIBRARY

Texts and Studies
in the History of Logic and Philosophy

Editors:

N. KRETZMANN (Cornell University)
G. NUCHELMANS (University of Leyden)
L. M. DE RIJK (University of Leyden)

1. M. T. Beonio-Brocchieri Fumagalli, *The Logic of Abelard*. Translated from the Italian. 1969, IX + 101 pp.
2. Gottfried Wilhelm Leibniz, *Philosophical Papers and Letters*. A selection translated and edited, with an introduction, by Leroy E. Loemker. 1969, XII + 736 pp.
3. Ernst Mally, *Logische Schriften*, ed. by Karl Wolf and Paul Weingartner. 1971, X + 340 pp.
4. Lewis White Beck (ed.), *Proceedings of the Third International Kant Congress.* 1972, XI + 718 pp.
5. Bernard Bolzano, *Theory of Science*, ed. by Jan Berg. 1973, XV + 398 pp.
6. J. M. E. Moravcsik (ed.), *Patterns in Plato's Thought. Papers Arising Out of the 1971 West Coast Greek Philosophy Conference.* 1973, VIII + 212 pp.
7. Nabil Shehaby, *The Propositional Logic of Avicenna: A Translation from al-Shifā: al-Qiyās*, with Introduction, Commentary and Glossary. 1973, XIII + 296 pp.
8. Desmond Paul Henry, *Commentary on De Grammatico: The Historical-Logical Dimensions of a Dialogue of St. Anselm's.* 1974, IX + 345 pp.
9. John Corcoran, *Ancient Logic and Its Modern Interpretations.* 1974, X + 208 pp.
10. E. M. Barth, *The Logic of the Articles in Traditional Philosophy.* 1974, XXVII + 533 pp.
11. Jaakko Hintikka, *Knowledge and the Known. Historical Perspectives in Epistemology.* 1974, XII + 243 pp.
12. E. J. Ashworth, *Language and Logic in the Post-Medieval Period.* 1974, XIII + 304 pp.
13. Aristotle, *The Nicomachean Ethics.* Translated with Commentaries and Glossary by Hypocrates G. Apostle. 1975, XXI + 372 pp.
14. R. M. Dancy, *Sense and Contradiction: A Study in Aristotle.* 1975, XII + 184 pp.
15. Wilbur Richard Knorr, *The Evolution of the Euclidean Elements. A Study of the Theory of Incommensurable Magnitudes and Its Significance for Early Greek Geometry.* 1975, IX + 374 pp.
16. Augustine, *De Dialectica.* Translated with Introduction and Notes by B. Darrell Jackson. 1975, XI + 151 pp.